RIVERS OF POWER

ALSO BY LAURENCE C. SMITH

The World in 2050:
Four Forces Shaping Civilization's Northern Future

RIVERS OF POWER

HOW A NATURAL FORCE RAISED KINGDOMS, DESTROYED CIVILIZATIONS, AND SHAPES OUR WORLD

LAURENCE C. SMITH

Little, Brown Spark
New York Boston London

Little, Brown Spark
Hachette Book Group
1290 Avenue of the Americas, New York, NY 10104
littlebrownspark.com

First Edition: April 2020

Little, Brown Spark is an imprint of Little, Brown and Company, a division of Hachette Book Group, Inc. The Little, Brown Spark name and logo are trademarks of Hachette Book Group, Inc.

Maps by Matt Zebrowski

ISBN 978-0-316-41200-1 (hc) / 978-0-316-49716-9 (int'l ed)
LCCN 2019951484

10 9 8 7 6 5 4 3 2 1

LSC-C

Printed in the United States of America

For the astounding Selma Astrid
and her powerful currents

CONTENTS

RIVERS OF POWER

INTRODUCTION

When the first rains came, the world changed forever.

They would have come sooner, by 100 million years or so, if not for the collision with another planet. It was roughly the size of Mars. The crash was so massive that our young Earth was engulfed in fire and mostly melted. A giant piece sheared off and most likely became the Moon. A magma ocean churned and raged on the surface of the wrecked planet.

Then, the primordial surface began to cool. A crust of iron-rich rocks hardened on the Earth's magma ocean. A lighter crust formed too, floating like slag in a smelter. A smattering of zircons, best known today for their use as inexpensive gemstones, began to crystallize. Trace remnants of them can still be found today in ancient rocks of Australia, Canada, and Greenland.

Australia's zircons have been dated as far back as 4.4 billion years. This signifies that continental crust began forming on Earth much earlier than previously thought, perhaps just 200 million years after our planet first congealed out of a swirling disk of cosmic dust and gas some 4.6 billion years ago. The chemical composition of these crystals tells us that at least trace amounts of liquid water were already present, despite Earth's extreme volcanism and the inferno caused by its collision with the other young planet. Like miniature time machines, zircons offer a glimpse into the Earth's earliest eons, the Hadean (named after Hades, the Greek god of the underworld) and the Archean (derived from the Greek word *arkhē*, meaning "beginning"). From their chemistry, we've learned that our world's early magma ocean cooled quickly and that continents and water came soon after.

By four billion years ago, if not sooner, rain began falling from the young sky. Water pooled in lakes and seeped into the ground. Water flowed overland into rivulets, streams, and rivers to the newly filling seas. Water evaporated into the poisonous air, condensed into clouds, and rained down again to complete the cycle. Water began eroding Earth's infantile, thickening continental crust, opening an eternal war against the continents.

Bit by bit, the rains broke down the high ground and filled in the low. They dissolved rock and loosened minerals. They weathered mountains and nudged their detritus downhill. Trickles found one another, came together, and grew stronger. They joined together, over and over, until millions combined into a powerful force—rivers.

The rivers had one job to do: Move it all downhill. Down, down to the sea.

Where tectonic collisions raised mountains, water and gravity allied to grind them down. Where plates wrenched open new seas, rivers toiled to fill them. Muddied with silt, their waters coalesced like roots toward a stem. Gravel jostled and rolled down flowing branches to some final destination.

At the end of their run, the rivers died in seas and lakes. Spent, they dumped their sediments and evaporated like spirits, rising aloft, back to the high ground to attack, flatten, carry, and dump again. Mountains are tough, but even the mightiest spire is doomed to fall to this tireless enemy. The water cycle outlasts them all.

By at least 3.7 billion years ago, rivers were steadily depositing sediments into the world's oceans. A couple hundred million years later, blue-green cyanobacteria—the world's earliest photosynthesizers—began generating whiffs of oxygenated air. Around 2.1 billion years ago this oxygen production surged. Pyrite (fool's gold) and other readily oxidized minerals disappeared from the riverbeds. The world's iron-rich soils reddened like rust.

Another billion-plus years passed. Then, between 800 and 550 million years ago, the ocean's exhalation of oxygen surged again. Sponges, flatworms, and other strange new marine lifeforms appeared. In the eons to follow, these early organisms would persevere, advance, and eventually populate our world in strange and marvelous ways.

Meanwhile, the continents thickened and crashed about. New mountain ranges swelled and were beaten down. But their rocky substance was transmuted, not lost. The relentless rivers spread their debris across the lowlands, building wide, flat valley plains. Deep stratigraphic sequences were laid down, layer upon layer, slowly infilling the basins and seas. River deltas pushed out fingers of new land far offshore into the oceans.

Rivers are literally universal. From orbiting spacecraft, we see them on other worlds. Mars once had abundant liquid water, and its surface is now scarred with the dry channels, deltas, and layered sedimentary deposits of ancient rivers. At this very moment there are rivers flowing actively on Titan, a cold distant moon of Saturn. Their fluid is liquid methane and the bedrock they carve is believed to be made of ice, but the valleys, deltas, and seas they are busily creating are eerily Earthlike in pattern and form.

Oceans opened and closed. Continents crashed and bulged. Some of the rivers' sediments were dragged deep down into the Earth's mantle on the backs of subsiding tectonic plates, where they were ferociously squeezed and heated. The cooked remains further thickened the continents and rose, like heated wax in a lava lamp, to cool into the hardened roots of new mountain ranges. Eventually some of this same material was exhumed, pulverized, and carried off by rivers once again, for yet another journey back to the sea.

Our world's destructive construction project never ends. Mountain ranges rise, then are pounded into sand. Their debris fans out across river valleys, deltas, and offshore continental shelves. Every earthquake, every landslide, every raging flood, marks just another little rumble in this ceaseless war between two ancient forces—plate tectonics and water—that are locked in combat for the shape of our world's surface. Their war will continue for at least another 2.8 billion years or so, until our dying, expanding Sun boils away every last drop into steam.

Today, the rivers struggle to carry their loads to the sea. They slide past hardened cities, yoked by dams, throttled by engineers, overlooked by most. Still, the rivers prevail. They will outlast us all.

But we will not endure without them.

~

The many ways that humans use rivers have varied by region and changed over time. Yet their importance to us has persisted because they provide us five fundamental benefits: access, natural capital, territory, well-being, and a means of projecting power. The manifestations of these benefits have changed, but our underlying needs for them have not.

In Egypt, for example, the Nile River once supplied natural capital in the form of silt-rich floodwaters. Today, it supplies natural capital in the forms of hydroelectricity, municipal water supply, and high-value riverfront real estate in downtown Cairo. The Hudson River once supplied the Lenape with fish, then European immigrants a transportation gateway to the continent. Today this same river provides access to precious waterfront parks in New York City, a teeming metropolis with little greenspace. The details varied, but the five overarching benefits remained durable. Through these provisions, rivers have been serving human civilizations ever since our first great societies rose along the banks of the Tigris-Euphrates, Indus, Nile, and Yellow Rivers in present-day Iraq, India-Pakistan, Egypt, and China.

Throughout human history, our attraction to rivers has been on display through art, religion, culture, and literature. They meander through the paintings of Van Gogh and Renoir, the writings of Muir and Thoreau, the music of Johann Strauss II and Bruce Springsteen. Enduring works of fiction, from Twain's *Huckleberry Finn* to Coppola's *Apocalypse Now* rose from the dark waters of their imagining. People the world over are calmed by the sounds of rushing water from brooks, fountains, and sleep-therapy machines. To bathe in the Ganges River is a moving religious moment for millions of Hindus, as is the baptism rite for millions of evangelical Christians. Virtually all of our great cities—the world's epicenters of knowledge, culture, and power—have a river running through their core.

This book contends that rivers hold a grandly underappreciated importance to human civilization as we know it. Rivers are of course important in many practical ways: they supply us with drinking water, coolant for power plants, and a means of removing sewage, for example. But they also shape us powerfully in less visible ways. Our repeated explorations and colonizations of the world's continents were guided by rivers. Wars, politics, and social demographics have been jolted by their devastating floods. Rivers define and transcend international borders, forcing co-operations between nations. We need them to produce energy and food. The territorial claims of nations, their cultural and economic ties to each other, and the migrations and histories of people trace back to rivers, river valleys, and the topographic divides they carve upon the world.

Rivers are beautiful, but their hold over us is far more than aesthetic. Their allure stems from the intimate relationship we have shared with these natural landscape features since prehistoric times. Our reliance on them—for natural capital, access, territory, well-being, and power—has sustained us for millennia and grips us still.

CHAPTER 1

THE PALERMO STONE

Near busy downtown Cairo, at the end of a built-over island, stands a modest square structure. Atop its thick stone walls is a cone-shaped parapet. Surrounding the structure are a small palace, a museum celebrating the famed Arabic singer Om Kolthoum, and the Nile River.

If you enter the structure, you'll discover it caps a stone-walled shaft, nearly forty feet square, sunk deep into the earth. Stone steps descend in flights around its walls. Rising up from the gloom, in the center of the space, is a massive marble column. Deep marks are cut into its octagonal sides at roughly equidistant intervals. Along the chamber's lower walls three subterranean tunnels radiate away toward the Nile.

Inside this chamber, the din of the Middle East's most crowded city is hushed. The entire shaft is invisibly encased in concrete, and the tunnels are sealed. But if they were reopened, the waters of the Nile River would rush in, flooding the chamber until its water level matched that of the river outside. The marks on the column could then be used to measure the height of the river. And for five thousand years this device, and dozens of others like it, served a critical purpose for the governance and survival of human civilization in Egypt.

The structures are called *Nilometers* (in Arabic, *miqyas*), and they were designed to empower Egypt's rulers with constant updates on the progress of the annual flood of the Nile, one of the most predictable rivers in the world. Each summer, under rainless skies and baking heat, the Nile River would mysteriously swell over a period of weeks, crest its banks, gently flood the land, and then slowly recede. To an ancient people living in what is today called the Sahara Desert, this unfathomable, glorious annual event was miraculous and divine. They understood nothing of the physical reasons for the predictable arrival of the annual flood, but everything about its power.

The importance of the Nile River flood to the early Egyptians cannot be overstated. By enabling them to raise food and livestock in the desert, the flood enabled their civilization to exist. It will therefore not surprise you that accurate knowledge of the exact day and water level of its crest was of paramount importance to Egyptian rulers. Under their watchful eyes, the river's water level would crawl up the Nilometer, halt, and then start a gradual descent, thus marking the passage of that year's maximum water supply (see color plate). Pronouncements would be uttered, criers would shout, and slaves would rush to breach temporary earthen dams, allowing the Nile's water to spill into the parched fields. Under the glaring sun, it spread out across the valley, inundating it for a few weeks before receding. Farmers followed close behind, pushing seeds into the rich mud. Like a dark sash cutting through the desert, the low-lying areas along the Nile River valley and its lobe-shaped delta jutting into the Mediterranean Sea turned green. The crops were irrigated and another year of survival was assured.

Even before the season's first seed was planted, Egypt's rulers already knew how big the harvest would be. They knew whether the following year would bring festivals or famine. The height of the flood peak, marked by the Nilometers, correlated directly with how much acreage of surrounding land would be inundated and planted. They already knew how much grain the farmers would produce—and they proclaimed the year's *taxes* accordingly.

~

The Nilometer on Roda Island, in present-day Cairo, was founded in 861 CE, making it one of the youngest in Egypt. Earlier Nilometers were built along ancient, now-vanished river channels for thousands of years. At least four types have been discovered: a simple stone column, a wall or corridor of steps descending down into the water, a circular walled well (often ringed by steps descending down its sides) with openings to the river, and a combined well and column like the one in Cairo. The cut markings measured units of length called *cubits*, roughly as long a man's forearm. In what may be the first direct linking of a quantitative scientific measurement to public health, Pliny the Elder used data from a Nilometer in Memphis, a now-ruined city on the Nile River delta, to predict the food security of ordinary Egyptians. Twelve cubits, he wrote, spelled death from famine. Thirteen, hunger. Fourteen, gladness. Fifteen, all was good. Sixteen, unbounded joy!

For thousands of years, Egyptians (and later, their invaders) used Nilometers to track the progress of the annual Nile River flood. So critical were these measurements that the annual water levels were etched, together with other important records like agricultural yields and tax revenues, onto an important stone slab (a *stele*) known as the Royal Annals. Seven fragments of the Royal Annals are held today by museums in Cairo, London, and Palermo. Their significance went unrealized for decades because they were untranslated and mostly bought from random antiquities dealers. One chunk is said to have been discovered while in use as a doorsill. The largest and best-preserved piece lay ignored until 1895, when a French visitor to Palermo noticed it languishing outdoors in the corner of a museum courtyard.

That fragment is now known as the Palermo Stone. Together with its six companions, it has shed more light on the history of ancient Egypt than any other archeological discovery. It was carved during the pharaonic Fifth Dynasty in the twenty-fifth century BCE, and includes a record of the Nile River's peak flood level for every year as far back as the early First Dynasty, approximately 3100 BCE. The river's flood history thus comprises humanity's longest written scientific data record. Researchers have since used it to illuminate everything from natural climate variability to occasional outbreaks of social upheaval in ancient Egypt.

In the early 1970s, the Harvard University astronomer Barbara Bell was the first to link low Nile flood levels to the so-called First Dark Age of early Egypt, when the long-stable civilization fell into anarchy, bringing its Sixth Dynasty and Old Kingdom to a crashing end. Some of the grimmest decades in Egyptian history are associated with this period, with an overall breakdown of social order that included revolution, murder, looting, tomb robbing, and farmers too frightened to plant the fields.

Such episodes were rare. To help prevent them, Egypt's rulers tightly restricted access to the information provided by their Nilometers. The structures were built inside or next to controlled temples, and only priests or other high-ranking officials were allowed to inspect them. The agricultural planning that was built around this system is one reason why the pharaonic empires survived for some three thousand years, suffering only three dark ages between the emergence of a unified Egyptian state (the First Dynasty, around 3100 BCE) and its conquest by Alexander the Great and subsequent absorption into the Roman Empire in 30 BCE.

It is unknown whether Egypt's seductive and last independent ruler, Cleopatra VII, reflected upon the importance of the Nilometers as she lay dying from poison, but they were nonetheless a lasting part of the pharaohs' legacy. Egypt became a vassal of the Roman Empire with the Nile River valley supplying about one third of Rome's grain supply. Cairo's Nilometer remained in operation until 1887, a thousand-year run. Flood-irrigation agriculture continued until 1970, when the construction of the Aswan High Dam ended the annual flood of the lower Nile valley. Egypt thus traded the Nile's natural capital of flood irrigation for that of stable, controlled irrigation and hydropower generation.

For millennia, the benevolent annual flood of the Nile River sustained the Egyptian people and entrenched the power of its rulers. Without it, one of the world's most stable and glorified civilizations would never have happened.

The Land Between the Rivers

Egypt's pharaonic dynasties were unusually persistent, but they were hardly the first river societies. By 4000 BCE—more than a thousand years before the first Egyptian pyramid was built—an ancient civilization of Sumerians established some of the world's oldest cities in lower Mesopotamia, the dry but fertile plains stretching between the Tigris and Euphrates Rivers south of Baghdad in modern-day Iraq. The origins of this civilization date back even further, perhaps as early as 7000 to 6000 BCE, when small-scale farmers began experimenting with stream irrigation in northern Iraq. The technologies they devised to divert water out of naturally flowing channels onto farmland would lead to humankind's lasting invention, the *city*.

Mesopotamia, which means "Land between the Rivers," was utterly different from Egypt. The Nile River flood spilled out gradually and gently upon the land and arrived in August, coinciding with peak agricultural water demand. The Tigris and Euphrates

Rivers, in contrast, flooded from March to May, too early for optimal planting. To be useful for irrigation, the water had to be stored behind check dams, cajoled into smaller, more tenuous planting areas, or lifted up out of the main channels later in the year, when water levels were low. Their floods were violent, unpredictable, and damaging. The Nile flowed sedately through a single, stable channel, whereas the Euphrates River, in particular, split into madcap shifting branches, sometimes abruptly abandoning its channels to carve out a new path. Such sudden course shifts, called *avulsions*, instantly rendered useless years of backbreaking work spent building dikes and irrigation ditches.

Mesopotamian farmers had no choice but to follow the rivers' gyrations around, digging new waterworks and cleaning out others choked with silt. Even without avulsions, the Land between the Rivers was prone to destructive floods that would periodically wipe out farmers' efforts and bury their fields in useless sand. The chronic damage wreaked by floods, avulsions, and ordinary sedimentation led to a shifting patchwork of high-maintenance planting fields and irrigation infrastructure that was intermittently constructed and abandoned.

Despite these problems, irrigating the fertile plain was highly productive. Farmers grew more food than they could consume, creating surpluses that could then be traded. The population grew, and as early as 5200 BCE, fledgling towns with names like Eridu and Uruk began appearing along the shifting riverbanks. The economic and political forces that drove their experiences are still debated, but one thing is certain: Without the food surpluses reaped from irrigation agriculture, these settlements would never have developed.

As the young towns expanded, agricultural production intensified, the irrigation waterworks became more complex, and water planning became more centralized. Decision-making power shifted to urban priests and bureaucrats, and crops were taxed to support a ruling class. Other technological advances—like the ox-drawn scratch plow and the invention of long, narrow fields (which require fewer turns to plow than square fields)—further accelerated the production of wheat and barley. Uruk, Eridu, and other settlements along the Tigris-Euphrates waterways grew into regional centers of power and formidable city-states. Trade blossomed, and the river channels themselves became vital transportation routes for boats. By 4000 BCE urbanization had spread throughout southern Mesopotamia, with some eighty percent of all Sumerians living in cities. Uruk, with an estimated population of as many as 100,000 people, was the largest city that the world had ever seen.

Sometime after 2000 BCE, Uruk's river channel shifted and abandoned the city. Devoid of water, its populace emptied out. Today, from satellite mapping, we can see how the locations of dozens of abandoned Sumerian cities and hundreds of archeological sites align with faint traces of ancient, long-dry river channels crisscrossing the Land between the Rivers. Uruk lies half-buried in blowing sand, its ghostly ruins marking the first of many empires—Akkadian, Babylonian, Assyrian, Ottoman, British, and Iraqi, to name a few—that would rise and fall over the following millennia in this deeply historic region.

Ark of the Tigris-Euphrates?

The twining Tigris-Euphrates river channels provided the food, water supply, and trade routes enabling Sumerian city-states to flourish. From these cities rose the beginnings of organized government, commerce, religion, and some of the world's oldest known literary writings.

These writings include a remarkable series of twelve clay tablets, inscribed in crisp cuneiform lettering, dug from the ancient ruins of Nineveh near present-day Mosul, Iraq. One of them tells the story of an ancient visionary who received divine instruction to build a gigantic ship. It had to be built large enough to accommodate representatives of all living animals. Later, a black cloud rose from the horizon and a catastrophic flood inundated the world for six days and seven nights, destroying everything except the great ship. When the floodwaters subsided, the visionary and his passengers found themselves perched on a mountaintop, but they were otherwise unharmed and free to repopulate the world.

This flood story sounds like the biblical Genesis account of Noah and the Flood, but in fact predates the Old Testament by more than a millennium. It is written on one of the twelve tablets of the Epic of Gilgamesh, which describe a legendary king of Uruk. The earliest stories on these tablets, which themselves date to 1200 BCE, trace back to 2100 BCE, and they were likely adapted and rewritten from even older versions.

From other archeological evidence we know there really was an Uruk king named Gilgamesh who ruled sometime between 2800 and 2500 BCE. His city is mentioned in the Old Testament (called "Erech," Genesis 10:10), and other similarities between the Epic of Gilgamesh and the biblical story of Noah and the Ark suggest that both documents share a common origin in ancient Iraq. Based on the age of the Sumerian tablets (and some even earlier writings from which they borrow) the origins of the Great Flood legend can be traced back for thousands of years, perhaps even into the Neolithic Period in Mesopotamia, around eight to twelve thousand years ago.

While there is absolutely zero geological evidence of a global flood during these (or any other) times, numerous credible studies suggest that a real-world local catastrophe may have inspired the legend. One popular hypothesis is that rising global sea levels forced a surge of seawater over the Bosporus outlet into the Black Sea. Another singles out the lowermost end of the ancient

Tigris-Euphrates river valley, now part of the seafloor of the Persian Gulf.

At the height of the last ice age (called the Last Glacial Maximum, or LGM) around 21,000 years ago, global sea level averaged some 125 meters lower than today. The present-day Persian Gulf, extending from Dubai to Kuwait City, was a broad river valley dotted with freshwater lakes. What is now the Strait of Hormuz, one of the most strategic and militarized maritime corridors in the world, was a flat, wide, fertile river valley.

Due to its gentle topographic slope, this ancient valley became inundated when global sea levels rose rapidly from approximately 10,000 to 4000 BCE, due to the melting of continental ice sheets and thermal expansion of ocean water as it warmed. The rising sea expanded more than a thousand kilometers inland, drowning the river valley and creating the present-day Persian Gulf. Because of the valley's extremely flat topographic relief, the sea's advance averaged more than 100 meters per year, and sometimes more than 1 kilometer per year.

To the region's human inhabitants, living on what is now the muddy seabed of the Persian Gulf, the relentless inundation of their homeland over the course of many generations was surely a noticed and traumatic event. Oral (and eventually written) accounts of their forced migration may have passed down to their descendants and could be the original source of the Gilgamesh epic, the Old Testament account of Noah and the Ark, and other ancient Great Flood legends.

Secrets of the Sarasvati

The Egyptian and Sumerian civilizations are highly studied, but they pale in sheer scale to the Harappan civilization of South Asia. Between approximately 2500 and 1900 BCE this remarkably advanced culture flourished along the Indus and

Ghaggar-Hakra River valleys and their tributaries across a large swath of modern-day Pakistan and northwestern India. One archeological study at Bhirrana, India, suggests that the Ghaggar-Hakra settlements may have been established even earlier, between 7000 and 5000 BCE. If true, this means the Harappan civilization emerged nearly two thousand years before the earliest Sumerian city-states.

Harappan villages, towns, and cities eventually spread across more than a million square kilometers, an area larger than the Egyptian and Mesopotamian civilizations combined, from the foothills of the Himalayas down to the coast of the Arabian Sea. Their inhabitants invented writing, granaries, brick-lined water wells, and urban planning. They built sophisticated municipal plumbing systems with running-water baths, toilets, aqueducts, and enclosed sewers—some *two millennia* before these same defining features of modernity appeared in ancient Rome.

Like the Egyptians and Sumerians, the Harappans were a river people. They planted and irrigated crops of wheat, barley, millet, and dates in the fertile floodplain silt deposits. Food surpluses supported urbanites living in well-planned cities built with burnt bricks. The best-studied are Mohenjo-daro and Harappa, two particularly large cities being excavated in modern-day Pakistan. In the mid-nineteenth century, their ruins were mined by colonial British railroad engineers, who used the ancient bricks for track ballast. The antiquity and importance of these ruins were not appreciated until 1924, when the first archeological studies began.

For reasons not fully understood, the Harappan civilization and its remarkably advanced technologies faded away. People vanished first and most precipitously from the Ghaggar-Hakra River valleys around 1900 BCE. Out of nearly fifteen hundred currently known Harappan archeological sites, almost two-thirds are located along the ancient, dried-up remains of the Ghaggar River and its tributaries. A leading hypothesis for the disappearance is that a prolonged weakening of the Indian monsoon season caused these rivers to dry up, making the region too arid for crops. Today, satellite images reveal numerous traces of long-gone river channels in this now-parched area. All that remains of the Ghaggar River is an erratic, intermittently flowing ribbon that disappears in the Thar Desert. Its demise might even have inspired the mythical disappearance of the Sarasvati River, first mentioned in the oldest known Sanskrit religious writings, the Rigveda, around 1500 BCE, and an important legend in India today.

Yu the Great Returns

Farther to the east, Chinese civilization took root in the fertile but deadly floodplains of the Yangtze River (Chang Jiang) and Yellow River (Huang He). As early as 6000 BCE, rudimentary rice cultivation appeared in two separate locations along the Yangtze, near the present-day cities of Shanghai and Changsha. China's earliest discovered rice paddies, found at the Kuahuqiao archeological site near present-day Hangzhou, appeared around 5700 BCE. The early Majiabang and Hemudu cultures depended upon freshwater foods like wild rice, lotus seed, cattail plants, and fish.

To the north, in the Yellow River valley, a group of millet farmers called the Yangshao culture flourished from about 5000 to 3000 BCE. China's earliest known writings come from this region, inscribed first on bones, then on bronze, wood, bamboo, and eventually paper. According to these inscriptions, Chinese civilization began along the Yellow River, with the San Huang (Three Sovereigns), the Wu Di (Five Emperors), and the first dynasties—Xia, Shang, and Zhou (Three Dynasties). Chinese oral tradition holds that the first Xia dynasty originated sometime between 2200 and 2070 BCE and that its founder was Yu the Great.

Yu the Great is an important traditional figure in China. According to the *Shiji* (Records of the Grand Historian), a series of massive floods once wreaked successive waves of disaster upon the millet farmers living in the Yellow River valley. For nine years, Yu's father tried unsuccessfully to block the floods by building dams and dikes. But Yu succeeded where his father failed, by cutting canals to drain away the water instead. After thirteen more years of digging ditches, often toiling alongside his laborers, Yu the Great mastered the Yellow River and won the loyalty of his people. His political power consolidated, he created Xia, the first dynasty of China, declaring himself its first king with all subsequent kings to be determined by hereditary succession.

This traditional history still enjoys widespread popular acceptance in China but has clashed with evidence-based archeology. In the 1920s a skeptical group of historians formed the Yigu Pai ("Doubting Antiquity School") to challenge the existence of both Yu the Great and the first Xia dynasty. In particular, they pointed out that the purported start of the Xia dynasty (around 2200 to 2070 BCE) fails to align with any noteworthy expansion or innovation in archeological artifacts unearthed in the Yellow River valley. The Erlitou culture—associated with a wave of innovations in ceramics, bronze, and jade—did appear later (as early as 1900 BCE) but its timing was at least two centuries later than the traditionally held start of the legendary Xia dynasty.

Nearly a century later, scientific advances in the mapping and dating of ancient flood deposits may help to settle the matter. In 2016 a team led by Qinglong Wu of Peking University published in the journal *Science* provocative geological evidence of a cataclysmic flood in the upper reaches of the Yellow River. Their research suggests that an earthquake triggered a landslide far upstream in the Jishi Gorge, a deep canyon cut by the upper part of the river near the Tibetan Plateau. The landslide buried the gorge with nearly 800 feet of debris, creating a natural dam that blocked the river. An enormous lake began infilling behind the dam, eventually overtopping and bursting it. The lake emptied out, sending a catastrophic flood raging through the Yellow River valley. Radiocarbon dating of the flood deposits fix the date of this great flood at 1922 ± 28 BCE.

That date aligns perfectly with the dawn of the Erlitou culture, some 1,500 miles downstream of Jishi Gorge, where the Yellow River leaped from its channel and began carving out a new course across the North China Plain. Subsequent flows passing through this newly created riverbed would have taken many years to equilibrate and control, possibly corresponding to the legend of Yu the Great's (and his father's) years of effort to control the river's course. In the general vicinity and time of this avulsion appeared an explosion of technological innovation in ceramics, bronze, and jade. This coincident timing supports the idea that China's civilization rose from the aftermath of a cataclysmic flood in the Yellow River valley, with the Erlitou culture in fact being the "lost" Xia dynasty.

We may never learn if Yu the Great was a person or myth, and further research is needed to confirm or dispute this particular study. But this ancient legend draws a clear connection between successful flood control along the Yellow River and the organization of mass labor, top-down political power, and the origin of political dynasties. Put another way, a society's recovery from an extreme river flood may well have sparked nearly four thousand years' rule under a dynastic system of government in China.

Wittfogel's Waterworld

A common theme runs through the four stories of great civilizations described thus far. All formed along wide, flat river valleys having fertile silt soils but scanty rainfall. Rain-fed agriculture was difficult or impossible to sustain in these areas, so river irrigation was a critical requirement for these societies' growth and survival.

The natural capital of rivers—water for irrigation, fertile floodplain soils for planting—was exploited and managed through ingenious human inventions like Nilometers, canals, dikes, dams, and water-lifting devices such as Archimedes' screw. Despite omnipresent dangers of floods, avulsions, and droughts, agriculture was so successful that food surpluses—in particular storable grain—became possible. Taxing and trading these surpluses enabled the emergence of new professions, social classes, and cities.

Freed from the daily chore of producing their own food, people developed new occupations, including scribes, accountants, priests, merchants, politicians, and soldiers. They concentrated in compact settlements where it was easier to interact and which could be fortified against marauders. And as these settlements grew, inventors found new ways to press rivers into service—for municipal water supply, sewerage, and trade with other population centers.

As these societies grew in number and complexity, so also did their need for agricultural productivity. Indeed, the very survival and political stability of these great civilizations depended upon good maintenance of their irrigation waterworks. So imperative was this requirement that the German-American historian Karl Wittfogel (1896–1988) named them "hydraulic societies."

A survivor of Hitler's concentration camps, Wittfogel immigrated to the United States after international pressure persuaded the Gestapo to release him. He became a naturalized American citizen and academic, serving on the faculty of Columbia University and later the University of Washington. In part due to his searing experience with "that inferno of total terror," (as he described the Nazi camps), Wittfogel became consumed with understanding the origins and nature of totalitarian power.

Today, he is mainly recalled for two things: icily testifying before two McCarthy-era investigative committees that several of his fellow scholars were probably Communists, and writing a deeply influential book titled *Oriental Despotism: A Comparative Study of Total Power* (1957).

Together with a series of academic articles, *Oriental Despotism* argued that the level of organization and mass labor needed to sustain river irrigation infrastructure—and therefore food surpluses, taxation, and a ruling class—was so great that the rise of an authoritarian, bureaucratic society was the most likely political outcome in such places. A ruling class run by priests or kings, argued Wittfogel, would inevitably emerge to impose top-down control over the water, land, workers, flood control, and repairs necessary to manage large-scale complex waterworks and keep them in good working order. State control of this infrastructure would inevitably lead to an oppressive, bureaucratic government too formidable to be resisted by ordinary citizens.

Hydraulic societies were thus stable, but their very survival relied upon continued good management of their waterworks, thus encouraging strong authoritarian rule and state control. Weak management or catastrophic failure—from neglect, war, avulsions or climate change—could lead to food shortages and political upheaval, tipping them into decline or collapse. Examples of such disruptions permeate the histories of ancient Mesopotamia, the Indus River valley, and China, and happened occasionally even in the relatively stable Nile River valley.

Oriental Despotism expressed a groundbreaking idea that has since evolved after decades of study and debate. Not all historical civilizations with complex waterworks evolved into authoritarian states, for instance. There are also known examples in which an authoritarian state came first and developed waterworks later. And environmental dependencies, even critical ones like water and food systems, do not in themselves determine political outcomes. But even Wittfogel's critics agree that a core requirement for the establishment of the early great civilizations along the Nile, Tigris-Euphrates, Indus, and Yellow rivers was the successful exploitation of these rivers' natural capital and an ability to adapt or recover from disruptive floods and avulsions. Where these requirements were achieved, food surpluses, taxation, and social hierarchy followed. Manipulation and control of rivers thus led to the birth of densely populated, complex, hierarchical societies (authoritarian or otherwise). The dawn of the multi-profession, multi-class city ruled by elites had begun.

Knowledge, from the Breasts of Hapi

One of the many strengths of a taxed, occupationally diverse urban society is that it can afford to support a few thinkers. What problems, what questions consumed the earliest intellectuals? Few would deny the benefits that science, engineering, and law have brought to humanity, or that these problem-solving approaches broadly underpin our world today. Where did these three distinctly human institutions come from?

Although they would not become recognizable in modern form until at least the Renaissance, the origins of science, engineering, and law trace back to the earliest civilizations, and often involved appropriating natural capital and human well-being from streams, rivers, and other forms of flowing water. Around 3000 BCE, an unknown artist carved a depiction of an irrigation canal into the stone mace-head of "King Scorpion," a mysterious pre-dynastic ruler in lower Egypt. Through trial and error, the Sumerians, Harappans, Egyptians, and Chinese diverted rivers away from their settlements and onto their croplands with canals and dikes. Terracotta pipelines and sewers were used extensively by the Hellenic civilization of ancient Greece and later copied by the ancient Romans. The Romans plumbed their public baths, fountains, and villas with pipes of lead or baked clay, and built extensive aqueduct systems to divert water to their cities. In the first century BCE, Vitruvius of Rome dedicated an entire volume of his famous treatise *De architectura* to issues surrounding the diversion and management of flowing water.

Pragmatic and uninformed, this early body of work was feeling its way through the basic tenets of what we today call civil engineering. Some of these early proto-engineers' accomplishments were spectacularly impressive, such as the Romans' famous arched aqueducts. Many of these gravity-fed structures are still standing today. Yet in other ways the ancient proto-engineers were surprisingly ignorant.

Take, for example, one of the most basic of all river measurements, discharge (also called streamflow). *Discharge* describes the volume of water passing a fixed location over some unit interval of time (e.g., gallons per minute, cubic meters per second, cubic kilometers per year, and so on). It is used to govern everything from the management of dams and reservoirs to the maximum legal flow rate of the showerhead in your bathroom. The discharge of a river, canal, or aqueduct is equal to the planar cross-sectional area of its flow multiplied by the average flow velocity across that plane. It's a simple concept, but the ancient Greeks and Romans believed that enlarging or constricting a watercourse alone controlled its rate of flow. They strangely overlooked or ignored the importance of flow velocity, which can be regulated by adjusting a watercourse's slope.

A rare exception was a mathematician and proto-engineer named Hero (also called Heron) who lived in Alexandria, Egypt, during the first century CE. He is famous today for writing two radical books, *Pneumatica*, which set forth some founding principles of hydraulics, and *Dioptra*, which basically invented land-surveying. The content of these books was so groundbreaking that Hero has since been called the First Engineer. Among numerous watery concepts presented in *Pneumatica* (the siphon, irrigation, and drainage, for example) Hero explained how flow *velocity,* not just cross-sectional area, was needed to correctly determine flow through an aqueduct, river, or spring. His lucid description of the necessary calculations was roundly ignored by his contemporaries, and the concept of discharge would not be instituted for another sixteen centuries. A Benedictine monk and student of Galileo Galilei named Benedetto Castelli finally established it for good in his work *Della Misura dell'Acque Correnti* (On the Measurement of Running Waters) in 1628.

It is an easy trap to over-romanticize the wisdom of the ancients. Ancient Greeks favored vague, even stirring explanations of the natural world, but were remiss or generally uninterested in quantitatively measuring it. In his 1970 book *History of Hydrology,* Asit Biswas observes that even Aristotle, one of the greatest intellects of all time, propagated myths such as the assertion that men have more teeth than women, yet never bothered to confirm it by examining the mouth of either his wife or mistress.

The ancient Greeks' penchant for qualitative answers to questions is understandable, given how little people understood about the natural world in their time. And of the many mysterious natural phenomena begging for explanation, few so captivated the early philosophers as the movement of the stars in the night sky and the origin of the Nile River flood.

Thales, from Miletus (an important city in its time, with ruins still visible in what is today western Turkey), was the first to seek natural, rather than supernatural, explanations for the latter phenomenon. This was unheard of at the time. The ancient Egyptians, who had been living comfortably along the Nile's riverbanks for three millennia, believed that its life-giving annual flood poured forth from the heavy swinging breasts of the god Hapi, depicted in ancient carvings as an androgynous figure with beard, loincloth, and bulging, possibly pregnant, belly. Thales rejected this divine explanation, proposing instead that the south-blowing winds that appear in Egyptian summers pressed back against the river's north-flowing water, restraining it until the wind was overcome and a flood was released.

Thales' hypothesis was dismissed by Herodotus, who observed that the flood appeared even when the winds did not, and that other rivers experiencing winds in the opposite direction of their currents were unaffected. Herodotus in turn proffered his own baffling physical explanation, in which the flood was created by the seasonal movement of the Sun and Egypt's absence of rain. Over the next six centuries, many other Greek and Roman philosophers—including Diogenes, Democritus, Ephorus, Strabo, Lucretius, and Pliny—all proposed their own physical explanations for the origins of the river's flood. Although no actual field studies or measurements were attempted, the discourse marked the earliest glimmerings of what we would today call a scientific debate.

All of them were wrong. None understood that the seasonal precipitation cycle far upstream in the Ethiopian highlands drove the mysterious phenomenon. But these intellectuals' debates over the source of the Nile River flood—together with a handful of other debates in astronomy, cosmology, and mathematics—created a new style of proposing and debating rational, physically based hypotheses to explain the world around them. A rebuffing of mysticism and the pursuit of knowledge for its own sake had begun. The origins of science and rational thought can be traced as far back as Thales and early philosophical debates about the origin of the annual Nile River flood.

The Hammurabi Code

Since earliest times, societies have used rules to regulate civil order and the distribution of natural resources. If a rule is broken, justice is served by punishing the offender, compensating the victim, or both. Indeed, our innate hunger for some sort of justice system traces back at least four thousand years, when the first known laws were codified in writing. What were the earliest proto-lawyers concerned about, and how did their efforts influence our modern legal systems?

The first known written laws come from the archeological sites of Nippur, Ur, and Sippar, three ancient Sumerian cities along the banks of the Euphrates River. From the ruins of these cities were unearthed four small cuneiform tablets. One was deciphered in 1954, and cross-referencing and translating it with the others took nearly three decades more. As the painstaking linguistic research neared completion, it became apparent that the tablets, dated to around 2100 BCE, were the oldest legal texts ever found. Of at least thirty-nine clear laws inscribed on these ancient tablets, thirty-two have been deciphered. They are known as the Ur-Nammu Law Code.

A more extensive set of up to 282 laws, carved into a massive black stone slab standing over two meters tall, came into force three centuries later. The slab was dug from the ruins of Susa, about 250 kilometers east of the Tigris River in present-day Iran, but had originally stood in a Mesopotamian temple. It is called the Code of Hammurabi because it was placed there by Hammurabi, a powerful Babylonian king who ruled Mesopotamia from 1792 to 1750 BCE. Like the Ur-Nammu tablets, the Code of Hammurabi spelled out rules and punishments deemed necessary to preserve civil order and govern resources in one of humanity's earliest civilizations. It stands today in the Louvre and, together with the Ur-Nammu Code, provides a rare glimpse into the values of the civilization growing four thousand years ago in the fertile floodplain between the Tigris and Euphrates Rivers.

A reading of the Ur-Nammu and Hammurabi codes reveals that people of that time were worried about sex, violence, divorce, slaves, lying, and irrigation water. During Ur-Nammu times, most punishments were meted out as fines. For example, if a virgin slave girl was "deflowered with violence," the perpetrator would have to pay five silver shekels as punishment. The capital punishment of death was reserved for murder, robbery, or having sex with someone's (non-slave) virgin wife.

The Code of Hammurabi, being far longer, details many more crimes and punishments. Levels of punishment differed between social classes (nobility, free men, and slaves), and it expresses the earliest known codification of the *lex talionis* ("law of retaliation") concept of punishment:

> If a man put out the eye of another man, his eye shall be put out. If he break another man's bone, his bone shall be broken. If a man knock out the teeth of his equal, his teeth shall be knocked out.

This idea of justice as commensurate vengeance ("an eye for an eye, a tooth for a tooth") would go on to permeate the Hebrew Bible and Christian Old Testament and still lingers in parts of the world today. The concept of differing levels of punishment for the same crime—depending on the perpetrator's social class—would reappear again in colonial territories and in America during the slavery era and still persists less visibly today.

These earliest legal codes also mandated a surprising number of protections for vulnerable members of society. For example, in the Code of Hammurabi a virgin female rape victim would be held blameless, and her rapist executed. If a male slave married a free woman, the slave's master could not enslave their children. Victims of robbery would be reimbursed by the government. Some of these ideas seem curiously progressive for such ancient times and, in some parts of the world, would be seen as progressive even today.

What about governance of natural resources? In both of these codes, river water (or the plants they irrigated) was the main natural resource cited. Related crimes included failing to properly maintain ditches and dams, accidentally flooding a neighbor's field, and stealing irrigation equipment. The earliest known laws thus set forth legal precedents for responsible stewardship of water, personal liability, and property rights, alongside a sordid social brew of infidelities, sex crimes, assaults, thefts, loan defaults, graft, and other uncannily familiar human crimes.

Of course, some of the other legal concepts in these texts are utterly alien to us now. For example, the power to adjudicate was sometimes turned over to nature. If a man was suspected of sorcery or a woman of infidelity, the Euphrates River served as both judge and executioner: Accused individuals were thrown into the water and their guilt or innocence determined by whether they survived or drowned.

Rivers for All

Today, it is a bedrock legal principle around the globe that rivers cannot be owned. Even in countries with strong capitalist traditions, such as the United States and the United Kingdom, rivers are a class apart, reserved for the public good. This puts rivers in a category distinctly different from that of most other natural resources. It is extremely common for land, trees, minerals, and water from other natural sources (e.g., springs, ponds, aquifers) to be deemed private property. Rivers, air, and oceans, however, are treated very differently. Where did this precedent come from, and how does it shape the legal systems of today?

The idea dates at least as far back as ancient Rome. According to the *Digest*, a major compilation of Roman legal writings commissioned by the emperor Justinian in 530 CE, early Roman lawyers established many legal principles surrounding the consumption of rivers, public access to rivers, and the rights of private landholders living along rivers. These writings reveal that Roman society strongly believed that unlike other freshwater sources, a permanent, flowing river (*flumen*) belonged to the public (*flumen publicum*). Romans were also concerned about preserving freedom of navigation, particularly the free passage of boats. Springs, intermittently flowing streams, groundwater wells, and other smaller sources of water could be privately owned, but any natural river that flowed year-round, even non-navigable ones, were owned by the public for the benefit of all. With stops and starts, this principle has largely endured, enabling public access and free passage of navigation on large rivers to this day.

Roman jurists also codified rights for riparian landowners. While any citizen could boat, swim, or fish in a river, *accessing* it was a different matter. To cross private land, a sort of public easement or right-of-way (called a *servitus*) had to be negotiated and balanced against the needs of the private landowner. It was up to land surveyors and judges to handle a *servitus*, much as public easements are adjudicated in the United States and other countries today.

The Roman government had authority to approve diversions, dams, and other major river projects. If necessary, it could even appropriate private land for this purpose, an ancient precedent of eminent domain. However, preserving the "natural flow" of a river was legally enshrined, meaning that riparian landowners were protected against another entity polluting or diverting a river's discharge upstream of their properties.

These three core ideas—freedom of navigation, public property *(res publica)*, and private property *(res privata)*—were disseminated across the vast Roman Empire. The first two established the freedoms of trade, communications, and travel along the Tiber, Po, Rhine, Danube, Rhône, Saône, Guadiana, Guadalquivir, Ebro, Orontes, and Maeander Rivers, in what are today Italy, Germany, France, Switzerland, the Netherlands, Romania, Hungary, Serbia, Bulgaria, Slovakia, Ukraine, Moldova, Spain, Portugal, Lebanon, Syria, and Turkey.

The idea of *res privata* (together with influence from English Common Law, from the Middle Ages) would ultimately evolve into the legal concept of water rights, which holds that private property owners along a river have a right to use its water. Centuries later (after being watered down by U.S. courts to relax the Roman requirement of "natural flow" to one of "reasonable use," thus allowing pollution by mills) this principle would enable an explosion of river-based industry and thousands of new settlements along millraces of western Europe and eastern North America. Had the Romans settled upon a different idea to govern rivers—say *flumen privata* instead of *flumen publicum*—the world would be a very different place today.

Wheels of Power

My life, and likely yours, calls for a fair amount of travel. If my destination is more than four hours away, I fly; if less, I drive. Only in the most remote locations where I work—places like Alaska, northern Canada, and Siberia—do I travel by boat. But where there are few airports and no roads, a canoe feels like a motorcycle. An aluminum boat with an outboard motor feels like an SUV. Rivers become roads, snaking through the wilderness.

Well-traveled game trails follow the banks. Even in the depths of winter, northern rivers are active transport corridors for people and animals, cruising down the smooth frozen surfaces as they have since primeval times. Traveling these remote water routes today feels adventurous and quaint. But until quite recently, following rivers was the primary way that people traveled and explored the interiors of continents.

Wolf prints along the Yukon River, Alaska. Since time immemorial, wildlife and people, including the earliest known human arrivals in North America, have used rivers and river valleys as natural travelways. *(Laurence C. Smith)*

Boat travel is relatively easy and has been around for millennia. No one knows when the first boat was made, and we probably never will. Indeed, they have been invented and reinvented many times throughout human history. The earliest models—dugout log canoes, bundles of reeds lashed together, wood frames skinned with bark or hide—have been unearthed in archeological sites all over the world. An eight-thousand-year-old log dugout was excavated at Kuahuqiao, the same archeological site near Hangzhou where China's earliest rice paddies appeared. Other ancient watercraft have been unearthed in Egypt, Mesopotamia, West Africa, Southeast Asia, India, the Americas, and Europe.

The planked boat, a technological breakthrough that increased seaworthiness, was invented at least as far back as 1670 BCE in England (a single surviving plank was discovered in 1996) and probably longer, and by at least 500 CE in California (by the coastal Chumash people). In the ninth and tenth centuries CE, Vikings used nimble planked warships to terrorize Europe, sailing up rivers to raid what is now the Baltics and Russia and founding new settlements in such places as present-day Normandy, Swansea, and Dublin. They colonized Greenland and the previously uninhabited island of Iceland, and explored the rocky coasts of northeastern North America a half millennium before Christopher Columbus arrived in 1492.

In the eleventh and twelfth centuries, cities like Antwerp, Ghent, and Rotterdam, fueled by shipping and commerce, rose along the navigable river channels of the Rhine-Meuse-Scheldt delta in Western Europe. Other important river towns included present-day Amsterdam, Florence, Paris, and London. The trade networks they established presaged an urban, mercantilist European economy and eventual doom for the old agrarian feudal system of manor houses and serfdom.

Europe's urbanization was further stimulated by growing

interest in using waterwheels to extract mechanical power from rivers. Since at least Roman times, small waterwheel mills had been used to grind grain into bread flour; they were commonplace in villages and manor estates throughout Europe. The simplest involved diverting a flowing stream onto a horizontal paddlewheel, turning a vertical crankshaft that had a millstone affixed to its top. The shaft was typically covered by a small wooden shed raised above the stream, to shelter the miller and grain as the slowly revolving millstone ground against a stationary one. This humble device had no gearing or flywheels, but was enormously important for grinding modest amounts of flour just about anywhere a reliably flowing stream could be found. Importantly, it was also used to grind malted barley for making beer.

By the eleventh century if not sooner, advances in waterwheel technology had led to construction of more powerful systems. An undershot waterwheel used gearing and a larger, vertically oriented wheel that was partially lowered into the water. For rivers too large to dam, huge undershot waterwheels were mounted on ships—essentially floating factory mills—that anchored by the hundreds in the Rhine, Elbe, and Danube Rivers. The medieval cities of Vienna, Budapest, Strasbourg, Mainz, and Lyons all relied on these ship mills at various times in their history. During the twelfth century, the city of Toulouse had at least sixty ship mills anchored in the Garonne River, and in Paris nearly a hundred churned in the Seine.

An overshot waterwheel also used a vertical wheel but was turned by an elevated stream of water, usually obtained by damming up a millpond or using a natural waterfall to divert a steady flow of water onto its top. This modification added the weight of falling water to the wheel's rotation, roughly doubling its power and requiring less water.

Through these technological advances, human exploitation

of river power expanded far beyond its modest origins of grinding grain into flour. Waterwheels began powering sawmills, paper mills, and iron foundries. They pumped water out of mines, spun wood lathes, and pounded felted cloth. Innovations in mechanical engineering proliferated as people figured out how to use gears, pulleys, flywheels, camshafts, pistons, and conveyer belts to optimize the conversion of streamflow to mechanical power. New industries rose alongside Europe's rivers, tapping water as well as energy. Paper and textile mills, in particular, required large volumes of river water to process wood pulp, cloth, and dyes and to flush away the waste. By the mid-eighteenth century, the mechanical world stood poised to enter the Industrial Revolution.

Valleys of the New World

Meanwhile, across the Atlantic, people had been traveling and living along North American rivers since the continent's earliest colonization by humans.

In 2019, a breakthrough archeological discovery found on the banks of Idaho's lower Salmon River (which flows into the Snake and Columbia Rivers) was announced. Dozens of stone spearpoints and blades, a hearth, and pits containing stone artifacts and animal bones were found. Radiocarbon dating of charcoal and bone from the site, now known as Cooper's Ferry, reveals that people started using it regularly around 16,000 years ago—the oldest radiocarbon-dated evidence of humans in North America. Very significantly, this timing predates the opening of an ice-free land corridor through the Cordilleran ice sheet (~14,800 years ago) by more than a millennium. North America's earliest known people, therefore, did not travel from the Bering Strait area over land, as once thought. Instead they came by sea even before the last ice age ended, traveling down the

Pacific coast and then turning east up the Columbia River, the first major river valley south of a massive ice sheet that covered western Canada and the Pacific Northwest at that time.

Between the eighth and fifteenth centuries CE, an advanced civilization emerged in the Mississippi River valley, culminating in the construction of a capital city called Cahokia, on a now-abandoned river bend near the confluence of the Mississippi and Missouri Rivers near present-day St. Louis. Together with a series of smaller population centers along the valley, it radiated power throughout the region.

Its people farmed, crafted fine artworks, and erected enormous pyramids made of earth and wood. They developed new customs, ideas, and political systems. At its culmination of influence, around one thousand years ago, Cahokia had administrative and religious centers and a political culture spreading across a large fraction of the continent. Six centuries later, its descendants would greet the Spanish explorer Hernando de Soto; later still they would battle the westward expansion of white immigrant settlers. Remnants of the capital city's pyramids can still be seen in the Cahokia Mounds State Historic Site, just a few minutes' drive from downtown St. Louis.

Looking south to Central America, recent archeological discoveries reveal that the mysterious Maya civilization was a type of hydraulic society. Using laser-scanning technology to penetrate the thick forest canopy, a team of researchers led by Tulane University discovered more than sixty thousand ancient structures in northern Guatemala, including houses, palaces, ceremonial centers, and pyramids. This huge civilization, with an estimated population of 7 to 11 million people, was supported by more than a thousand square kilometers of intensively cultivated farmland irrigated by ditches, terraces, canals, and reservoirs tapping the headwater streams of what is today called the Rio San Pedro.

When European explorers and fur traders arrived in northeastern North America, they encountered thriving aboriginal societies using lightweight, durable canoes of curved wooden ribs sheathed in birch bark. The fur trade found the design so ingenious that French-Canadian *voyageurs* adapted larger versions of these canoes to push deep into Canada from the 1690s to 1850s. The largest versions (called *canots du maître*) were basically freight vessels, up to thirty-six feet long and six feet wide at their centers. Singing lustily and chomping pemmican, the vigorous *voyageurs* used the rivers like freeways, paddling and portaging eighteen hours a day to empty pelts out of the continent.

By the peak of the fur trade in the early 1800s, the *voyageurs* numbered perhaps three thousand men. Romanticized in their time and ours, they were in reality a small group of low-wage contract workers whose lives on the rivers were astonishingly hard. Most were illiterate and couldn't swim. The grueling work curved their spines and deformed their feet. They died from drowning, starvation, and accidents. They were porters, not trappers, delivering company goods to aboriginal people while adopting aspects of indigenous clothing and ways of life. In their time, the *voyageurs* were the long-haul truckers of the fur trade.

They also spread death. DNA analysis of samples from tuberculosis victims across central and western Canada has traced the pathogen back to a single lineage of *Mycobacterium tuberculosis*. Beginning in about 1710, this bacterium was carried by a few *voyageurs*, perhaps just two or three individuals, along the river trade routes deep into Canada, where it persisted at a low level for more than a century before subsequently expanding in the late nineteenth and early twentieth centuries, when conditions favoring the pathogen's spread promoted deadly tuberculosis epidemics. The spatial drainage pattern of rivers thus shaped not only where early colonists explored, traded, and settled, but also seeded the early tuberculosis epidemiology of Canada.

The *voyageurs* opened up northern North America to foreign trade from the Great Lakes and Western Canada all the way up into the Arctic. Their grueling canoe routes began in Montreal or Quebec City, following the St. Lawrence and Ottawa Rivers to headwaters deep inland. These routes were used by fur companies—Hudson's Bay Company and the North West Company—to consolidate their territory, profits, and regional power. The trading posts and garrisons built along the routes established the first permanent foreign settlements in interior North America.

South of Canada, competition between France and Britain for control of one of North America's most vital waterways has a long history. France's aspirations for the Mississippi River dated at least as far back as the explorations of René-Robert Cavelier, Sieur de La Salle, who in 1682 was the first European to sail down the entirety of its length all the way to the Gulf of Mexico. He laid an engraved plate at its mouth and claimed the entire Mississippi River watershed for France, naming it Louisiana in honor of King Louis XIV.

La Salle had no way of knowing it, but the size of his claim was vast, covering some 3.2 million square kilometers and draining what is now thirty-one U.S. states and two Canadian provinces. France did little with the remote territory, but by 1749 competition from British colonists had intensified. The Ohio River Valley, the great eastern fork of the Mississippi River that makes up much of its watershed, was also claimed by Britain, the Virginia Colony, and the aboriginal Iroquois confederation. A group of Virginian land investors had formed a private venture called the

Ohio Company, seeking from the king of England a title to 500,000 acres of the upper Ohio River Valley (today western Pennsylvania), in order to survey and then sell the land to settlers. The king agreed, granting 200,000 acres up front, plus a promise of 300,000 more if the company settled at least a hundred families within seven years and built a fort to defend them from the French. The Ohio Company's shareholders included Robert Dinwiddie, the lieutenant governor of the Virginia Colony, and Lawrence Washington, half brother of a young Virginian named George Washington.

George Washington's Big America

It is rather astonishing how much of the ethos of American business culture today can be traced back to this one perseverant man. Young George Washington would one day become an intrepid Revolutionary War leader and the first president of the United States, but he never really hungered for political fame or wartime adventures. Deep down, what he loved was real estate.

George Washington wanted to own land. A lot of it. He was so interested in land that as a teenager he began a career as a land surveyor. He saved his earnings, and at the age of eighteen began buying choice parcels that he found through his work. His preferred targets were large tracts of river bottomland, which he prized for their flat, fertile terrain and easy access by boat.

Washington was born into Virginia's elite planter class, so it was not unusual that he viewed land accumulation as a path to fortune. What set him apart was his fixation on land *west* of the Appalachian Mountains, far from the thirteen British colonies established on the continent's eastern Tidewater seaboard.

He was especially intrigued by the Ohio River Valley, which by 1749 his half-brother Lawrence was seeking to develop through the Ohio Company. Together with the Virginia Colony's claim, this was too much for the French, who had considered the valley theirs since the time of La Salle. They dispatched from Montreal one Captain Pierre-Joseph de Céloron de Blainville, who journeyed down the Allegheny and Ohio Rivers with a force of more than two hundred men. He hung copies of the king of France's coat of arms on trees along the riverbanks and buried engraved lead tablets underneath to reassert France's claim. England and Virginia responded by launching their own expeditions including, in 1753, the twenty-one-year-old land surveyor George Washington.

Upon seeing the Ohio River firsthand, Washington immediately grasped its strategic importance as a gateway to the rest of the continent. He hustled back to Virginia and urged Lieutenant Governor Dinwiddie to build a fort at the "Forks of the Ohio," the point of land where the Allegheny and Monongahela Rivers come together to form the Ohio. Dinwiddie agreed, and a small fort named Fort Prince George was begun on the site.

Within a year the French captured it, built a bigger one in the same place, and named it Fort Duquesne. Dinwiddie sent Washington to deliver a letter ordering the French soldiers to leave. The order was rebuffed. Britain then escalated the situation, sending the decorated Major General Edward Braddock and two regiments of infantry to capture Fort Duquesne, again also sending young George Washington.

Marching four abreast in bright red uniforms, Braddock's redcoats were slaughtered by hidden French muskets blazing from behind the trees. Braddock was killed and his army suffered nearly a thousand casualties out of some fourteen hundred soldiers caught in the firefight. Young Washington survived and helped lead the retreat. In 1758 he returned to the Forks of the Ohio yet again, accompanying an even larger army of British soldiers and Virginia militiamen. This time the French retreated, burning and abandoning Fort Duquesne, and France lost Middle America forever.

The victors quickly replaced Fort Duquesne with Fort Pitt, an enormous pentagon that was large enough to house an entire British regiment. Traders, trappers, and land prospectors also settled on the sharp point of land aiming west like an arrow down the Ohio, toward the Gulf of Mexico. The settlement that grew around Fort Pitt would soon have an indelible name: Pittsburgh.

Within a few short years, Fort Pitt and its commanders would play important roles in the American Revolutionary War. George Washington would lead ragtag rebel armies against his former compatriots to an improbable American victory and become the new country's president. After the war, George Washington bought thousands of acres of land in the Ohio River Valley, and remained focused on its security and development for the rest of his life.

Pittsburgh became a launching pad for migrants seeking opportunities west, away from the Appalachians and the increasingly owned and expensive Tidewater states. European settlers floated down the Ohio, destined for homesteads in the remote Northwest Territory, which would eventually become the U.S. states of Ohio, Michigan, Indiana, Illinois, Wisconsin, and northeastern Minnesota.

George Washington's recognition of the strategic importance of the Ohio River broadened the aspirations of an otherwise small collection of British colonies. It influenced the thinking of Benjamin Franklin, John Adams, and John Jay—America's chief negotiators for the 1783 Treaty of Paris, which ended the Revolutionary War—who insisted that the western boundary of the newly independent United States should extend all the way to the Mississippi River. Two decades later, this same strategic thinking would be executed yet again by the third U.S. president, Thomas Jefferson, in his quest to capitalize on France's increasingly tenuous hold on the New World.

Jefferson oversaw the Louisiana Purchase, the greatest land deal in modern history. In 1803, after some initial overtures from Napoleon, the fledgling United States bought out all of France's remaining Louisiana Territory claims, plus the city of New Orleans, for just $15 million. This bought the United States control of the entire Mississippi River basin and—unknown to Napoleon and Jefferson at the time—more than doubled the size of the young country, including some of the most productive farmland on Earth. Relative to the young country's size at that time, a rough modern equivalent would be the present-day United States paying less than $500 million to purchase all of Canada.

Nearly a century would pass before America's aboriginal peoples were dispossessed, its political stability secured, and its domain mapped. But with the stroke of a pen, the United States gained official control of the biggest watershed in North America and its numerous huge arterial rivers. This would later allow ships to penetrate the continent and pass freely between its resource-rich interior and the outside world.

Even as negotiations with France were under way, Jefferson was dispatching river exploration campaigns, in hopes of projecting American power all the way to the Pacific Ocean. One such campaign was the famous Lewis and Clark Expedition, which Jefferson commissioned to voyage up the Missouri in search of "the most direct and practicable water communication across this continent." Meriwether Lewis departed from Pittsburgh in July 1803 in a wooden keelboat floating down the Ohio River; William Clark joined the expedition a few months later in Louisville, Kentucky. Over the next three years, they mapped many rivers, including the Mississippi, Missouri, Osage, Platte, Knife, Yellowstone, Salmon, Clearwater, Snake, and finally, the Columbia, which they journeyed all the way to its confluence with the Pacific Ocean.

If not for the aggressive pursuit of large inland rivers by Washington and Jefferson, colonial North America might very easily have been divided into a British Canada, a French Middle America, a Spanish West, and a small independent United Eastern States. Had that occurred, the world would be very different today. Many colonial territories did not win independence from their colonial masters until the 1950s and 1960s, and some as late as the 1980s. But through acquisition and exploration of the continent's major watersheds, Washington and Jefferson set into motion the idea and destiny of a massive single America spreading from the Atlantic to Pacific Oceans.

CHAPTER 2

ON THE BORDER

I was staring at a dark cleft in a sloping concrete wall about fifty yards away. A steel mesh fence, patched like old jeans, ran along the top of a second sloping wall on which I was standing. The two slopes met at the bottom, with a shallow trickle of river flowing between them. Dirt, bushes, and rags lined the flat bottom of the concrete channel.

The flicker of movement came again. There, in the shadowy orifice of the storm drain, I could make out the shape of a man. No, two men. One had a black shirt, making him harder to see. The other wore a blue T-shirt and shorts. They were studying my two companions and me from inside the trapezoidal opening, about halfway up the far sloping wall. Behind them, I could just make out a circle of a sewer pipe and a bundle of clothes. I had noticed the men only because I was photographing the hardened river channel of the Rio Grande.

I asked one of my escorts, a U.S. Border Patrol agent named Lorena Apodaca, to translate a question for me. She waved and smiled at the two men hiding in the sewer drain across the river. "*¡Hola! ¿Podemos tomar una fotografía?*" One shook his head tersely. No. The other cracked a grin and waved back at us gaily. I put away my camera. We all stared at each other over the river for a while, until the men grew tired of the game. They retreated back into the darkness of the sewer, waiting for us to leave.

They were watching for an opportunity measured in seconds, to attempt an unauthorized crossing of the U.S.-Mexico border. When it came, they would dash across the shallow trickle of the Rio Grande and clamber up the sloping concrete embankment where I was standing. From there, they would either slice through the mesh—its nickname is the "tortilla fence" because it is cut and repaired daily—or sprint through a narrow driveway opening in the fence nearby. Behind us idled two white U.S. Border Patrol SUVs, carefully watching this gap.

If they got past the tortilla fence and the waiting border agents, they would race toward a much taller steel barricade, set back about 250 feet from the river. Standing eighteen feet high, its mesh is too fine for human fingers to clasp. Each man would be carrying two screwdrivers to jam through the holes. With the pair of tools, they could scale the barricade and then tumble down the other side into downtown El Paso, Texas.

Timing was everything. The whole gambit would be seen by video cameras and infrared sensors mounted on high towers posted regularly along the watercourse. If the men made it across the river and over both barriers, they could vanish in seconds by simply shedding their outer layer of clothing and stepping into the crowd. Downtown El Paso was thronging with people of Mexican and Central American descent, just like Ciudad Juárez, its sister city across the river.

El Paso is a beautiful city tucked away in the westernmost corner of Texas. Sun-baked red mountains overlook its colorful low-slung skyline and that of Ciudad Juárez, the largest city in the Mexican state of Chihuahua. Together with the nearby city of Las Cruces, New Mexico, the El Paso metropolitan area is home to a million people. Including the population of Ciudad Juárez, some 2.3 million people live in this urban agglomeration straddling three U.S. and Mexican states and two sovereign nations, all separated from each other by the Rio Grande.

El Paso–Ciudad Juárez marks the junction where the 690-mile-long land border between the United States and Mexico ends and its 1,241-mile-long river border begins. North of this junction, the Rio Grande meanders down from the southern Rocky Mountains, creating a pleasant greenbelt of irrigated fields that briefly separates Texas from New Mexico. It is patrolled only by birds, farmers, and the occasional kayaker. But upon reaching the triple-point confluence of Texas, New Mexico, and Mexico, the river veers east and becomes a caged strip of concrete and steel (see color plate), entering duty as the heavily guarded international border separating the United States and Mexico for more than twelve hundred miles, all the way to the Gulf of Mexico.

This triple junction is just a few minutes' drive from downtown El Paso. A low dam impedes the Rio Grande's flow just upstream of the international border, diverting much of its water into a concrete sluice called the American Canal. You can stand on the Rio Grande's left bank (defined as the riverbank on the left side, as one faces downstream) in Texas, and see both Mexico and New Mexico on the other side. Where the two meet, not far from the water's edge, stands a tall white obelisk.

It is International Border Commission Monument No. 1, erected in 1855 and the first of 276 boundary markers demarcating the land border between the United States and Mexico westward from this point all the way to the Pacific Ocean. If you stand next to the monument and peer up at the mountains, you can just make out the next one perched high on a crag a couple of miles away. Like J.R.R. Tolkien's fire-lit signal beacons linking the Middle-earth cities of Rohan and Gondor, each successive border marker in this area is strategically positioned so as to be just visible to the next.

David Taylor, an artist and professor at the University of Arizona, spent seven years finding and photographing each of these largely forgotten boundary monuments. Many can no longer be touched from the American side. They are walled out, effectively ceded to Mexico by a steel barrier of American fencing set back a few feet from the border.

A dozen yards southeast of Monument No. 1 is Mexico's bank of the Rio Grande, strewn with picnic leavings and stalked by a lonely-looking egret. From somewhere downstream come the happy shrieks and splashing of swimming kids. An old adobe building stands nearby, which in 1911 headquartered the beginning of the Mexican Revolution. From it, Francisco Madero and Pancho Villa launched a brief war on Ciudad Juárez, then surprised everyone by winning and ultimately toppling the Mexican federal government. From the El Paso side of the river, Americans watched the coup unfold from hotel rooftops. Others waded across the border to support the revolutionaries with gifts of oranges and cash. A century ago, the Rio Grande was just an easily forded natural river, a convenient, unguarded way for two neighbors to define their jurisdictions. Today, it is one of the most fortified and deadly river borders on Earth.

Every day, thousands of people legally cross this heavily guarded international border separating El Paso from Ciudad Juárez. They pass back and forth on pedestrian and vehicle bridges spanning the concrete-lined Rio Grande and its moat-like shadow, the American Canal. Thousands live in one city and work in the other or have family members living on both sides of the river. More than four million legal pedestrian crossings wash back and forth across these bridges every year.

Death lurks under the shadows of this cheerful hubbub. It waits under the bridges and in the searing desert mountains west of town. On land, the ground is impregnated with sensors, barricaded by steel fencing, and crisscrossed with dirt roads patrolled by white U.S. Border Patrol SUVs. Storm sewers from both cities dump runoff into the Rio Grande and are used by migrants and drug smugglers to move covertly underground. Agent Apodaca told me how recent visits by Pope Francis to

Ciudad Juárez and then-president Barack Obama to El Paso had necessitated border patrol agents to wriggle through these sewers armed only with handguns, to clear and secure them.

The two men we were exchanging stares with had selected one of El Paso's least dangerous places to attempt a border crossing. Along this particular 1.5-mile-long stretch of the Rio Grande, most of the river's discharge was barreling underground, through a tunnel beneath our feet. Two hundred yards downstream the water re-emerged in the American Canal. Barely thirteen yards across, the canal's narrowness is deceptive, tempting people to scale its fencing and swim across. With powerful flows up to eighteen feet deep and moving twenty-five miles per hour, many succumb to its currents and die.

The American Canal is enclosed by sturdy chain-link fencing on both sides, hung with warning placards in Spanish. But the crossing attempts still come. Every few hundred meters along the canal I saw emergency boxes stocked with ropes and life preservers. In El Paso, U.S. Border Patrol agents receive swift-water rescue training and are required to use it often. My escorts said there had already been at least eight drownings and many rescues that year.

According to the International Organization for Migration (IOM), an intergovernmental organization that is compiling a worldwide database of migrant deaths called the Missing Migrants Project, the most common cause of migrant death is drowning. Most die in the Mediterranean Sea, packed into sketchy boats that capsize during the perilous journey from North Africa to Europe. Their bodies commonly wash up on Libya's beautiful sand beaches. On land, migrants drown crossing rivers. The Rio Grande (also called the Rio Bravo, its Mexican name, in the IOM database) is one of the most lethal borders in the world, with well over two hundred documented drownings since 2015 at the time of writing.

Another is the Naf River, separating Myanmar from Bangladesh. Its victims are the Rohingya, a minority group of Bengali Muslims massed in the northern part of Rakhine state in Myanmar, a staunchly Buddhist country. Myanmar views the Rohingya as foreigners and trespassers, because their progenitors moved into the area during British colonial rule. Since the 1960s, Rakhine Buddhists and Myanmar's central leadership have intermittently disenfranchised and violently expelled Rohingya from the country. In 2017 an unusually savage purge killed thousands of Rohingya, forcing almost 700,000 people across the Naf River into Bangladesh (see color plate). At least 173 people drowned that year alone.

Others die in the Evros River separating Turkey and Greece, the Tisza River at the border between Serbia and Hungary, and the Danube River separating Bulgaria and Romania. In the Limpopo River, a border between Zimbabwe and South Africa, migrants are killed by hippopotamuses and devoured by crocodiles.

From the U.S. Border Patrol agents, I heard words of sympathy for migrants and revulsion for "coyotes," or people-smugglers, who urge their ill-equipped customers across deadly swift waters and searing desert mountains. "They don't care if they die," one told me grimly. Her compassion was real. Yet thousands of people are so desperate that they will risk their lives to avoid her. Borders are a knotty place.

Blue Borders

After visiting Texas, I came away deeply intrigued by humanity's use of rivers as political borders. What I had seen felt very different from the ancient hydraulic societies, which used rivers to unite people, not divide them, and to consolidate power, not fragment it. Yet a close look at any world map shows that many nations today use rivers and their topographic watersheds to define their territorial limits.

Along with its fortified Rio Grande and Colorado River borders with Mexico, the United States separates itself from Canada by means of the Rainy, Pigeon, St. Mary's, St. Clair, Detroit, Niagara, St. Lawrence, Saint John, and St. Croix Rivers. Prior to the terrorist attacks on the World Trade Center and Pentagon on September 11, 2001, these rivers were largely unguarded and could easily be crossed by boat in summer and by walking or driving over their frozen surfaces in winter. Today, tower-mounted infrared video cameras guard the St. Clair and Niagara Rivers separating the states of Michigan and New York from the Canadian province of Ontario. In winter, smugglers and U.S. Border Patrol agents play cat and mouse on snowmobiles along the frozen white highway of the St. Lawrence River.

For some fifteen hundred miles, the Amur River (Heilong Jiang) and its tributaries the Argun and the Ussuri separate northeastern China and Russia's Far East. The Yalu River separates China from North Korea, with some striking contrasts in development between the two banks. Germany uses the Rhine, Danube, Inn, Neisse, and Oder Rivers for borders. Brazil, Paraguay, and Argentina are separated by the Paraná River; Portugal and Spain by the Douro River; England and Scotland by the River Tweed. And peering closer, at subnational scales, we see rivers and their topographic watershed divides being used to delimit provinces, states, counties, and townships all around the world.

To my surprise, little quantitative research on this intensive use of rivers as political borders has been done. I persuaded Sarah Popelka, at the time a talented UCLA undergraduate geography major, to use Geographic Information Systems (GIS) software to tackle the problem (as we will see in Chapter 8, global "big data" analyses are becoming easier thanks to a proliferation of new satellite and geospatial datasets). By fusing political border and population density data with a high-resolution map of the world's rivers obtained from satellite remote sensing, we

The Yalu River separates two different worlds of development between North Korea (left) and China (right). *(Michal Huniewicz)*

created a new geospatial database of political river borders called Global Subnational River-Borders, or GSRB.

GSRB enables explicit identification and mapping of political borders set by large rivers at subnational and national scales. From this new database, we learned that at least 58,000 km (23 percent) of the world's interior (noncoastal) national borders, 188,000 km (17 percent) of the world's interior state/province borders, and 442,000 km (12 percent) of the world's interior county/local level political borders are large rivers. These numbers rise to a whopping 25, 20, and 22 percent, respectively, if only jurisdictions actually containing a large river are considered. In South America nearly half of all national borders are defined by rivers. Globally speaking, the number of neighboring pairs of political units sharing a river border number at least 219 country pairs, 2,267 state/province pairs, and 13,674 county/local pairs. These large numbers still underestimate the true use of rivers as political borders, because we did not include smaller rivers or topographic watershed divides in our study.

Put simply, our analysis quantifies the important role rivers serve in shaping the political jurisdictions in which we live and who our neighbors are. We will return to this original research again—in the context of global human population and the future of cities—in Chapter 9.

Lines of Expediency

Our penchant for using rivers and their topographic watershed divides to define political territories has a long history. Conquerors and empires often used these natural features as a convenient, highly visible way to delimit and negotiate territory. In medieval times, the Kingdom of France used the Saône, Rhône, Meuse, and Scheldt Rivers to define its own limits. As described in Chapter 1, the French explorer La Salle used the topographic watershed boundary of the Mississippi River—wherever it might happen to be—to claim a vast unexplored territory with no mapping or surveying whatsoever. He had no idea that his claim contained 1.2 million square miles, some 40 percent of the conterminous United States today, or that the geographical feature he used to define his claim would continue to be used as the land changed hands and was subdivided in the decades and centuries to follow.

American negotiators would later use the Mississippi River as an easily articulated, easily understood territorial goal in their negotiations with Britain to end the Revolutionary War. Two decades later, Thomas Jefferson and Napoleon Bonaparte used the same river, plus the western part of La Salle's original watershed boundary, to negotiate the uncharted Louisiana Purchase, a bloodless territorial triumph for the young United States that later proved to be one of the greatest land acquisitions in history.

In a poorly charted world, rivers and their topographic watershed boundaries make expedient natural delineators for ill-informed foreign conquerors to define and swap territory. After all, rivers are explicit, continuous, and long. Unlike a land survey, which requires both time and expense, they are free and already in place. They offer clear, objective targets for military conquest and for treaty negotiations. And aside from these cadastral conveniences, the rivers themselves provide access for exploration and trade, as well as natural capital in the form of bottomland timber, fertile soils, fish, and sometimes even gold. Militarily, rivers offer both a ready means of transporting personnel and supplies to remote theatres and an impediment to advancing armies. For all of the above reasons—territory, access, natural capital, and military power—rivers became a go-to natural feature shaping the exploration, military strategies, and territorial definitions of poorly charted continents by remote colonial powers.

The territorial expansion of the young United States exemplifies how rivers and their topographic watershed divides were commonly used as political borders prior to the availability of reliable maps. Beginning with Britain's Royal Proclamation of 1763 (which used the topographic divide between the Mississippi River basin and east-flowing Appalachian headwaters to delineate its Tidewater colonies), either the Mississippi River proper or its divides were used in negotiations for the Treaty of Paris (ending the Revolutionary War), the Louisiana Purchase, and the Oregon Treaty. Other rivers figured prominently in the annexation of Texas, as well as the creation of many U.S. states and of a sizable chunk of the Pacific Northwest. Throughout the eighteenth and nineteenth centuries, rivers and their watershed divides figured importantly in shaping the expansion of America's borders and political power (see map).

For example, today's U.S.-Mexico border might have been Texas' Nueces River instead of the Rio Grande, and its gateway city Corpus Christi instead of El Paso, had President James Polk not provoked Mexico into a lopsided war by annexing Texas in 1845. With Texas secured, Polk positioned American troops south of the Nueces River, which Mexico considered to be its northern border with Texas, triggering the Mexican-American

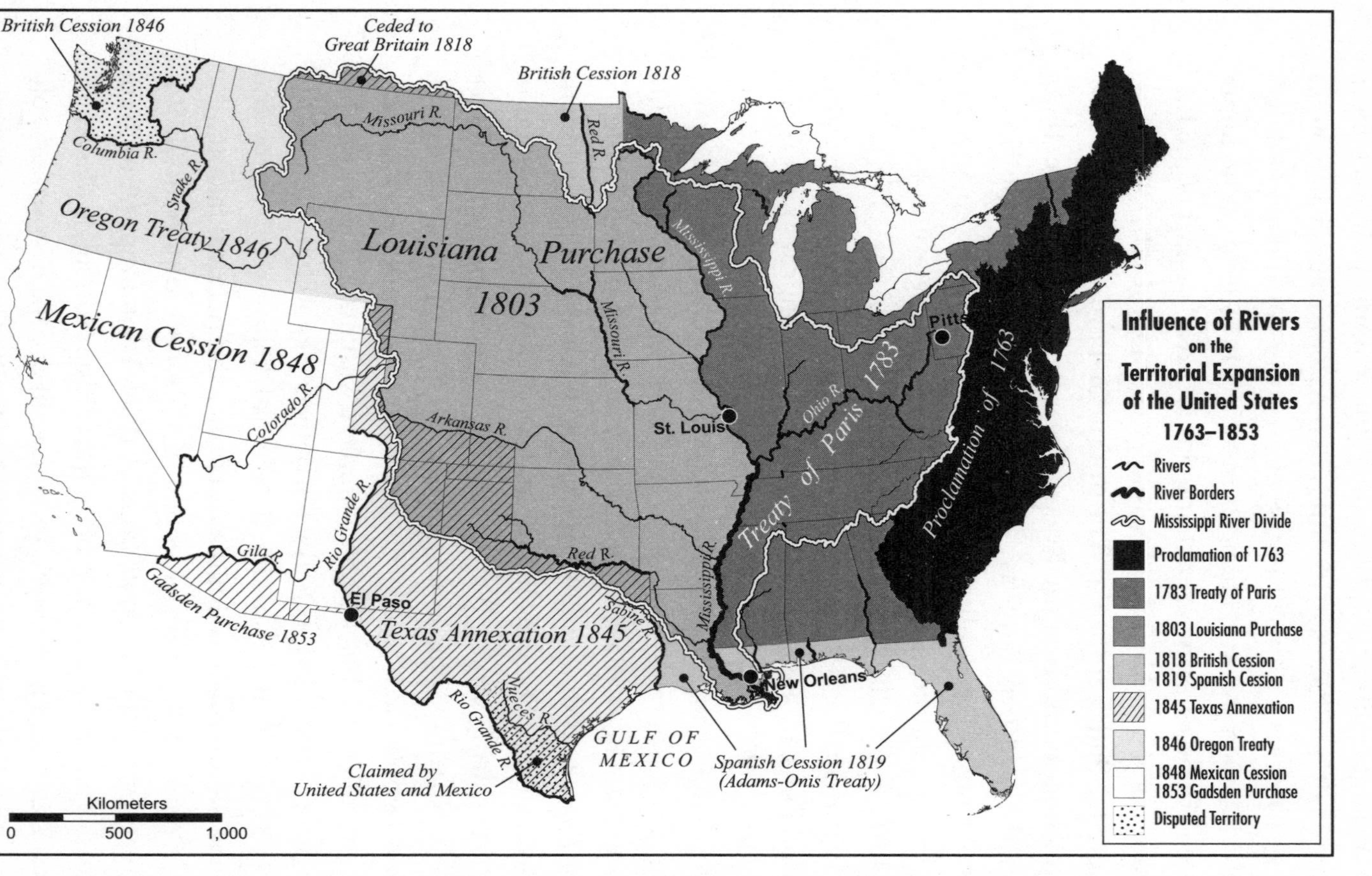

Colonial and national powers commonly used rivers and their topographic watershed divides to demarcate and barter land, as illustrated here by overlaying key rivers and river divides with important territorial expansion treaties of the conterminous United States between 1763 and 1853.

War. That conflict would cost the young country of Mexico more than half a million square miles of territory, roughly halving its size. Under the terms of the 1848 Treaty of Guadalupe Hidalgo, which ended the war, Mexico lost all territory west of the upper Rio Grande River, an area that today forms all or part of the U.S. states of New Mexico, Arizona, Colorado, Utah, Wyoming, Nevada, and California. The southern border of Texas hopscotched south, from the Nueces River to the Rio Grande. If not for the Gadsden Purchase of 1853–1854 (in which the United States bought an additional 29,670 square miles from Mexico), Arizona's southern international border today would be wavy, following the Gila River just south of Phoenix.

A fascinating exposé of how physical and social preexisting conditions influence the negotiation of political borders was written by Dr. Wesley Reisser, who worked in the U.S. State Department before attending graduate school in the Department of Geography at UCLA. His Ph.D. dissertation, later published as a book titled *The Black Book: Woodrow Wilson's Secret Plan for Peace,* tells of a little-known team of American geographers, historians, political scientists, and economists that convened in secret during World War I. The team was assembled by President Woodrow Wilson to concoct a master plan to redraw the world's political borders. The group prepared a top-secret document called the Black Book, complete with maps and plans, that Wilson carried daily into the Paris peace negotiations to end the war. These plans considered "natural features" of rivers and watersheds as well as linguistic, ethnic, and political factors.

Control of key rivers was one of Wilson's top priorities in his Treaty of Versailles negotiations. A Black Book proposal to grant headwaters of the Tigris and Euphrates Rivers to a single Mesopotamian state (today called Iraq) did not survive the negotiations, but others did. Many of the proposed borders considered access to rivers as a factor, sometimes even deliberately transferring their control to a favored country. Czechoslovakia (now the Czech Republic and Slovakia) was granted access to the Danube River at the city of Bratislava, for example. A controversial push to carve for Poland a "corridor to the sea" down the Vistula River to the German port city of Danzig (now the Polish city Gdańsk) was ultimately successful, despite the city's population being 90 percent German and the corridor effectively cutting Germany in two. Years later, after World War II, the U.S.S.R. expanded Poland's border farther west to the Oder and Neisse Rivers. In sum, certain rivers or their topographic watershed divides helped to shape the political geography of Europe and the Middle East after the first and second world wars. The legacies of these and many other historical conquests and treaties shaped by rivers have carried forward, affecting nearly a quarter of the world's inland political borders today.

The Size and Shape of Nations

The drawing of political borders, of course, is a purely human invention. They are contrived and negotiated by people, not physical geography.

Coastlines, rivers, and mountain ranges offer convenient natural features but compete with other considerations and aspirations. Nearly four thousand miles of ocean did not stop the United States from incorporating the islands of Hawai'i into its union. What, then, are some purely human factors that shape the size and number of nations?

In an important book called *The Size of Nations*, the political economists Alberto Alesina and Enrico Spolaore argue that economics, demography, and political freedoms limit how large a country can be. In general, a country's ideal size depends on tradeoffs between the advantages and disadvantages of having a large population. Advantages include things like having a larger economy, greater geopolitical clout in the world, and lower per-capita cost for infrastructure and public services. A large country can raise bigger armies and also has a dispersed pool of people from which to redistribute wealth, making it easier to weather local economic downturns and natural disasters.

However, large, populous countries are usually more heterogeneous, leading to diverse preferences, priorities, and cultures. Managing this diversity takes a toll on citizen satisfaction and therefore on a country's ability to govern, especially in open democracies. Failure to govern threatens civil order and the stability of the state.

Today, many large countries contain heterogeneous populations with diverse preferences that are challenging to sustain. To retain its current size, Iraq must negotiate the differing preferences of Sunni Muslims, Shia Muslims, and Kurds. Germany must manage anti-immigrant nationalists and liberal globalists. America presides over numerous malcontent factions, pitting rural conservatives against urban liberals and constituencies of varying socioeconomic status, gender, and race. Such countries survive at a cost of domestic rancor and political gridlock.

Former diverse nations that fractured under such pressures include the U.S.S.R., Czechoslovakia, the United Arab Republic, and Yugoslavia. Indeed, the late twentieth century was a time of extraordinary political fragmentation, when the number of sovereign countries in the world more than doubled. This political fragmentation was driven in part by increased democratization, which in turn increased secessions. Put simply, many countries decided to become smaller in order to give citizens what they wanted, even if it meant sacrificing the economic and geopolitical benefits of being big.

The Size of Nations helps to explain the social and economic forces that pressure nations to expand or break apart. But the final outcomes of these forces still play out upon the physical world. The utility of coastlines, topographic divides, and rivers as political borders was obvious to long-dead empire builders but is oddly ignored by academics today. These physical realities are not mentioned once in *The Size of Nations,* for example. Ask a political scientist why the world's political borders are where they are, and you will hear a fascinating chronicle of roles played by ethnicity, language, colonial history, religion, democracy, and authoritarianism. You will not hear about coastlines, rivers, or topographic watershed divides.

But as my research with Sarah Popelka, the map of America's territorial expansion, and a glance at any world map show, the physical world, too, shapes how political states choose to define themselves. People chart their political directions, but not in a vacuum. Alongside purely social forces, coastlines and rivers have influence. Rivers, and physical geography more generally, also contribute to the size and shapes of nations and thus the geospatial pattern of political power around the world.

Worries of Water Wars

These days, the preeminent role of rivers in international disputes is no longer the defining of borders. The more pressing issue is their water itself.

Despite falling fertility rates around much of the world, global population and developing-world incomes are still rising. To accommodate a richer, meat-eating planet of roughly 10 billion people—an approximate forecast of the global population in 2050—we must nearly double the world's current food production. Water conservation measures and technological advances such as crops genetically engineered to resist pests and disease will help us to meet this challenge. But in order to feed the growing number of people and livestock on the planet, we still need water. This promises further pressure on the world's already heavily subscribed rivers, streams, and groundwater aquifers.

It's easy to conjure up scary scenarios of where this might

lead. Indeed, an entire literature has sprung up around the threat of armed conflict over water. At the time of writing, a Google search on the term "water wars" yields a million internet hits and some thirteen hundred academic publications. Aaron Wolf, a geographer at Oregon State University who has long studied this topic, notes that water is "the only scarce resource for which there is no substitute, over which there is poorly developed international law, and the need for which is overwhelming." Three consecutive UN secretaries-general—Kofi Annan, Ban Ki-moon, and António Guterres—have openly worried that inadequate access to water could trigger civil unrest, mass migrations, and armed conflicts around the world.

Of especial concern are areas where populations are poor and international tensions are already high for other reasons. At least four major transboundary river systems fit this description. The Nile is now shared by eleven nations and nearly half a billion people. The Jordan River is shared by Israel, Jordan, Lebanon, the Palestinian territories, and Syria. The Tigris and Euphrates Rivers are shared by Turkey, Syria, Iraq, and Iran. The Indus is shared by Afghanistan, China, India, and Pakistan, with headwaters in the hotly contested mountainous region of Kashmir.

These rivers, already oversubscribed today, are essential for human survival—and shared by sworn enemies. The populations they support are growing and industrializing, with water demand rising. Might competition for control of these critical rivers lead to violence? Will the interstate wars of the twenty-first century be fought over water?

Mandela the Bomber

There is certainly a rational argument that they could. After all, Nelson Mandela, a globally revered champion of peace and social justice, felt compelled to kill over water.

Mandela emerged from twenty-seven years of political imprisonment to lead a progressive social movement, for which he earned a Nobel Peace Prize. He worked tirelessly to dismantle South Africa's racist apartheid system and eventually became the country's president in 1994.

Four years into Mandela's presidency, his South African National Defence Force (SANDF) raided the Kingdom of Lesotho, a tiny enclave nation in the highlands of southern Africa. Using attack helicopters and special forces, the SANDF wiped out a garrison of Lesotho soldiers guarding the Katse Dam, a newly built concrete arch dam and reservoir across the Malibamat'so River, that was part of an $8 billion joint water project between Lesotho and South Africa called the Lesotho Highlands Water Project. The Katse Dam was the first of five planned dams designed to impound runoff from the headwaters of the Orange/Senqu River and, via transfer tunnels, deliver some 2.2 billion cubic meters of water annually to Pretoria, Johannesburg, and Vereeniging, the industrial heartland of South Africa.

Sixteen Lesotho soldiers were killed when the SANDF soldiers captured the dam. Only after it was secured did this force proceed to Lesotho's capital city of Maseru to quell unrest over a contested election, the stated reason for the attack. Perhaps mindful of his peacemaker image, Mandela had the attack order issued by a subordinate named Mangosuthu Buthelezi, whom Mandela named as his acting president for a few days while he left the country for Washington, D.C., to receive the Congressional Medal of Honor from President Bill Clinton.

Academics and legal scholars have pored over the various arguments and treaties that South Africa later invoked as justification for the attack. The reasons given were found wanting. Quelling election protests simply did not add up as sufficient cause for South Africa to violate its commitments under the Southern African Development Community (SADC) Treaty and the United Nations Charter, for example. The real motivation for the attack appears to have been concern over threats to the Lesotho Highlands Water Project. At the time, it was one of the biggest river diversion schemes on the African subcontinent and a cornerstone of South Africa's long-term water security strategy.

Please take a moment to let that sink in: Nelson Mandela, a Nobel Peace Prize winner of great dignity, the same visionary who accepted prison rather than cease peacefully working to end apartheid, felt compelled to break international law by invading a sovereign nation over water. This may signify less about him than it does about the awesome importance of rivers. No president—not even Nelson Mandela—can defy the paramount importance of water security when weighing the well-being and national interests of his or her country.

Water Towers Make Water Wards

The reason Lesotho is so important to the well-being of South Africans is that its Maloti-Drakensberg Mountains form a "water tower," meaning a mountain range, typically surrounded by dry lowlands, that captures and funnels a very large amount of runoff into a major downstream river.

The downstream beneficiary in this particular case is the Orange/Senqu River, a huge artery flowing across the southern part of the African continent and a crucial water supply for South Africa. Other important water towers include the Ethiopian Highlands, which supply flow to the Blue Nile and Nile Rivers; the European Alps, which supply flow to the Danube, Po, Rhine, and Rhône Rivers; Africa's Bihé Highlands, which supply the Okavango and Zambezi Rivers; Central Asia's Pamir, Altai, Hindu Kush, and Tien Shan mountains, which supply the Amu Darya and Syr Darya Rivers; the Middle East's Taurus and Zagros mountains, which supply the Tigris and Euphrates Rivers; and America's Rocky Mountains, which supply the Colorado and the Rio Grande. The grandest water tower of all is the Tibetan Plateau and Himalayan Range, which form the headwaters of the mighty Indus, Ganges, Brahmaputra, Irrawaddy, Salween, Mekong, Yangtze, and Yellow Rivers, upon which nearly half of all living people depend.

Examining the locations of the world's water towers and the rivers they sustain, we see that many countries rely on water coming from someplace else. Lesotho controls a critical water tower for South Africa. Ethiopia controls a critical water tower for Sudan and Egypt. Angola controls a critical water tower for Namibia, Botswana, Zambia, Zimbabwe, and Mozambique. Nepal controls part of a critical water tower for India, and India part of one for Pakistan and Bangladesh. Turkey controls a critical water tower for Syria and Iraq. Tibet, Nepal, Bhutan, and Kashmir encircle a massive water tower essential to the survival of nine downstream countries and almost half the planet's population. By seizing Tibet in 1950, China gained control of a water tower critical not only for its own country but also for Bangladesh, Myanmar, Laos, Cambodia, Thailand, and Vietnam.

This arresting asymmetry between where river water is produced and where it is consumed has enormous power implications for political states. A country that controls a water tower—or the rivers emanating from it—wields a potentially existential threat over its downstream neighbors. The downstream country, called a "downstream riparian," worries that the upstream riparian may deplete or pollute the river's water before it crosses the border.

This hypothetical vulnerability is greatest if physiographic conditions actually permit an upstream riparian to divert or impound a river within its own borders. The United States, for example, wields immense upstream riparian power over Mexico because the Rio Grande and Colorado River flow long distances and attain large discharges within the United States prior to entering Mexico, with numerous attractive locations for diversions and dams. China and Laos wield similar power over Myanmar, Cambodia, Thailand, and Vietnam, with abundant potential locations to impound or divert the Mekong upstream of these neighbors.

Harmon's Folly

Hypothetically, an upstream riparian could consume or pollute every drop without regard for its neighbor, should it so choose. In 1895, Mexico's ambassador in Washington sent the U.S. secretary of state an urgent report expressing alarm that 500 miles of the Rio Grande along the U.S.-Mexico border had begun drying up completely in summer. Mexican farmers were abandoning fields they had been irrigating with the river's water for three hundred years. The population of Ciudad Juárez had halved. Why? Because far upstream of the border, new American diversions had been built to spread water from the upper Rio Grande across new croplands in Colorado and New Mexico. These diversions, wrote the ambassador, were killing off centuries of traditional water use along the Rio Grande riverbanks around Ciudad Juárez and El Paso.

U.S. attorney general Judson Harmon responded with a blistering legal opinion asserting, in effect, that the United States had no obligation to limit its domestic use of the Rio Grande. It had absolute sovereignty over its territory and therefore over any rivers flowing within its borders. The United States was free to do whatever it wanted with the Rio Grande's water resources upstream of the river's arrival at the U.S.-Mexico border.

The Harmon Doctrine, as it came to be known, exemplifies the existential fear of all vulnerable downstream riparians: the possibility that an upstream riparian might throttle its water supply. But in reality, most downstream riparians are also upstream riparians, and often with the same neighbor.

If a river is navigable, the downstream entity also holds a different kind of power, especially if its upstream neighbor is landlocked.

Despite being a downstream riparian, Germany still wields power over the Czech Republic because Czechs can't send ships down the Elbe River to the North Sea without passing through Germany. Upstream Hungary cannot use the Danube to reach the Black Sea without acquiescence from Croatia, Serbia, Romania, and Bulgaria. Unless Argentina concurs, ships from landlocked Paraguay cannot reach the Atlantic Ocean.

Under Harmon Doctrine logic, Mexico could justifiably respond by emptying the Rio Conchos, a major tributary of the lower Rio Grande that sustains its flow along the Texas border. Canada could impound the Columbia River inside British Columbia, prior to its entry to Washington and Oregon. Even the most strident U.S. nationalists quickly grasped that the Harmon Doctrine damaged its own goal of putting America first.

The next U.S. attorney general ignored the Harmon Doctrine. To halt any further diversion of Rio Grande water, the U.S. State Department took legal action to block a pending private dam project in New Mexico. Through a series of rulings, the U.S. Supreme Court affirmed their decision. The Harmon Doctrine was dead and the United States decided not to empty out the Rio Grande before it reached its neighbor.

Instead, the two countries negotiated a binding international treaty to fairly apportion the river's water between them. It entered into force in 1907 and was the first transboundary river treaty entered into by the United States. Its model of cooperation laid the foundation for how hundreds of other transboundary rivers are governed around the world today.

Like the 1907 U.S.-Mexico agreement, most are bilateral treaties. Some are between sworn enemies unable to find common ground on other issues. Since signing the 1960 Indus Water Treaty, India and Pakistan have aimed nuclear missiles at each other and fought three wars but have never violated the terms of their agreement to share the Indus equitably. During a protracted

Arab-Israeli conflict from 1979 to 1994, delegates from Israel and Jordan met clandestinely to work out a cooperative plan to share the Yarmouk River, a critical water source for the Jordan Valley, at a time when no formal diplomatic relations existed between the two countries.

Similar examples of cooperation between enemies date back as far as ancient Mesopotamia. Inscriptions carved into a Sumerian stone called the Stele of the Vultures (for its depiction of thousands of war dead being devoured by the scavenging birds) proclaimed an ancient pact between the Mesopotamian city-states of Lagash and Umma to equitably share the Tigris River after a bloody war. In 1804, the rival German and French empires agreed to share the Rhine River forever, a cooperation that has since been expanded to include Belgium, Switzerland, and the Netherlands. The origins of this further cooperation date back to the 1815 Congress of Vienna, which settled a new European order out of the wreckage of the Napoleonic Wars.

Today, there are nearly five hundred transboundary river-sharing agreements in force around the world, and their number continues to grow. Multinational treaties and commissions are in place for some of the world's most contested and oversubscribed river basins, including the Nile, Jordan, Tigris-Euphrates, and Indus Rivers, which are so often highlighted as flashpoints for armed conflict over water.

This decades-long trend hit yet another milestone in 2014, when Vietnam became the thirty-fifth country to join the UN Watercourses Convention (full name: Convention on the Law of the Non-Navigational Uses of International Watercourses), thus bringing to fruition a forty-four-year journey for this important piece of international law. To enter into force, at least thirty-five countries had to sign on.

The origins of the UN Watercourses Convention date to 1970, when the United Nations General Assembly voted to have the UN International Law Commission (ILC) draft a global framework for equitable sharing of rivers among nations. The ILC, in turn, turned to an even older body of work developed in the late 1950s and early 1960s by a scholarly group called the International Law Association (ILA). That effort culminated in a 1966 ILA conference in Helsinki, which issued the Helsinki Rules on the Uses of the Waters of International Rivers, a high-level set of rules for the shared governance of transboundary rivers. Importantly, the Helsinki Rules, as they came to be known, required riparian states to allow "reasonable and equitable" use of an international river by *all* riparian states along it. Today, this core idea is embodied in the UN Watercourses Convention and numerous transboundary river treaties around the world.

For other countries where transboundary agreements are either currently lacking or do not include all of a river's riparian states, the UN Watercourses Convention provides a useful starting point for negotiations. It also offers guidelines for other issues, such as pollution, which may not be covered by existing treaties. The UN Watercourses Convention formally entered force in 2014, and at the time of writing thirty-six nations were party to it, including the United Kingdom, Germany, France, Italy, Finland, numerous countries in Latin America, the Middle East, and Africa.

All Eyes on the Mekong

The story does not end there. As water resource demands grow and geopolitical powers shift, so too will the nature of cooperative multinational river-sharing arrangements.

The Mekong River, nearly 4,500 kilometers long, with a watershed of nearly 800,000 square kilometers, gathers itself in Tibet and then flows through China, Myanmar, Laos, Thailand, Cambodia, and Vietnam before entering the South China Sea. Its upper watershed, in China and Myanmar, is often called the Upper Mekong Basin, and its lower watershed, in Laos, Thailand, Cambodia, and Vietnam, the Lower Mekong Basin. In China the river is called Lancang Jiang, so its name is sometimes hybridized to Lancang-Mekong River. All of these different names and basin definitions for what is in reality one huge, continuous watershed underscore the divisive politics and visions surrounding this arterial river's future, especially in Southeast Asia, where the lower Mekong is currently one of the last great undammed rivers left on Earth.

The Mekong forms the backbone of Southeast Asia's food culture and supports local fishing and rice-farming economies throughout the region. Thailand, Vietnam, Laos, and Cambodia depend heavily on the river and its tributaries for agriculture, fish protein, and transport. Vietnam's Mekong River Delta and Thailand's Khorat Plateau are two of the region's most important growing regions, supplying roughly half of all rice produced in these two countries alone. Combined, the four countries produce some 60 million tons of rice annually. Two thirds of it is eaten domestically, feeding nearly 200 million people. The rest is traded, supplying roughly 40 percent of the global rice export market.

The Mekong and its tributaries are also a vital source of fish. Take, for example, an annual backwater flood in Cambodia's Tonlé Sap River, which creates one of the biggest freshwater fish baskets in the world. During a recent visit to Southeast Asia I was amazed at the ubiquity of freshwater fish in the Cambodian diet, from the busy markets and restaurants of Phnom Penh (situated at the confluence of the Tonlé Sap, Mekong, and Bassac Rivers) to rural villages surrounding Tonlé Sap Lake. The most famous of these are Tonlé Sap's "floating villages," which migrate back and forth, following the dynamic shoreline of this vast inland lake as it seasonally expands and then contracts.

Amid growing public debate about some proposed diversions and hydropower dams along the Mekong and its tributaries, Thailand, Vietnam, Laos, and Cambodia formed the Mekong River Commission (MRC) in 1995. The MRC is not so much a treaty as a functional governance structure, run jointly by its member nations. It has a high-level body made up of national ministers, a working council composed of agency department heads, and a secretariat's office to run things. The location of the secretariat moves around among the four nations, with offices so far in the capital cities of Bangkok, Phnom Penh, and Vientiane.

Development pressures along the Mekong are enormous, and the MRC has been a powerful arbiter of how that development takes place. While first and foremost a development commission, it is mandated to seek "wise use" of the river for the mutual benefit of all four countries and the well-being of their people. It seeks to identify and mitigate potential damage to agriculture, fishing, and local communities caused by river development, even across national borders.

A core job of the MRC is to identify and prioritize specific development projects, such as hydropower dams and irrigation diversions, in consultation with all four countries. It has a detailed process of notification, consultation, and consensus-building prior to recommending a major project for advancement. While member countries do not always abide by these procedures—Thailand, for example, recently ignored the MRC and briefly pumped water from the Mekong into its Huai Luang River during a severe drought—they are by and large followed. The MRC also supports water-quality monitoring, scientific research, specialized conferences, and educational outreach programs to monitor and manage the river basin in a transparent and comprehensive way.

Beginning in 2006, however, a serious challenge to the MRC's power was set in motion when Laos began planning two Mekong River dams, Xayaburi and Don Sahong, at the northern and southern ends of the country. These would be the first dams ever built in the Lower Mekong Basin. Laos' objective was (and is) to become the "battery of Southeast Asia," a major regional hydropower supplier, through electricity exports to Thailand and Cambodia. This would generate badly needed revenue in one of the world's poorest countries while also supplying a stable source of energy to its rapidly industrializing neighbors.

Laos' proposals generated angst among the other riparian nations, environmental groups, and international NGOs. They protested that the dams' impacts on rice farming, fisheries, natural ecosystems, and local communities were very poorly studied and potentially catastrophic. Laos proceeded anyway, submitting plans for a $3.8 billion dam at Xayaburi to the Mekong River Commission in 2010. The MRC determined that the impacts were indeed poorly understood and requested further scientific studies and data collection before proceeding. They also proposed a ten-year moratorium on any dams in the Lower Mekong Basin until the broader risks could be better determined.

Frustrated, Laos turned to unilateralism. Long-term energy contracts were signed with Xayaburi Power Company Limited, a private power company, and the Electricity Generating Authority of Thailand, the country's leading state-owned power utility. In 2012 Laos began construction of the Xayaburi Dam prior to a final decision by the MRC. A year later, the country submitted plans and impact assessments to the MRC for the second dam, at Don Sahong, but this time as a "notification," not an application.

Powerless, the MRC quickly kicked the matter upstairs to the uppermost levels of its member governments. High-level threats and diplomacy failed to stop Laos from signing more long-term energy contracts and starting construction on the dams. At the time of writing, after more battles and short-lived injunctions, both structures are nearly complete and a third, called the Pak Beng Dam, is poised to break ground. Planning for a fourth, called the Pak Lay Dam, is continuing despite the catastrophic 2018 collapse of an auxiliary dam of the Xe-Pian Xe-Namnoy hydropower project, in which forty people were killed and thousands displaced.

Laos' decision to unilaterally build dams on the Lower Mekong River laid bare the weakness of the Mekong River Commission. It has a limited mandate, few enforcement mechanisms, and no veto power over projects. Absent sharp teeth, the MRC's series of equivocal if well-intentioned calls for study and delay prompted Laos to abandon protocol and proceed on its own.

Surprised and disappointed, the nearly twenty international donors who fund much of the MRC—chief among them Finland, Australia, Sweden, Belgium, Denmark, and the European Union—slashed their funding from $25 million in 2015 to just $4 million by 2016 and 2017. If the MRC's own members won't abide by its rules, they reasoned, why fund it? After more than two decades of successful cooperative governance, the Mekong River Commission may well have met its match in the face of nationalist development pressures to build big dams in the Lower Mekong Basin.

~

The ultimate death sentence to the Mekong River Commission and its oversight of the lower basin's development may be signed not by Laos, but by China.

China is not an official member of the MRC, nor of ASEAN (Association of Southeast Asian Nations). It declined formal MRC membership in 1995 in favor of being a "dialogue partner," thus exempting itself from subjecting its own river development projects to MRC review. By 2018 China had built eight dams in the Upper Mekong Basin with at least twenty more under construction or planned. This activity has significantly altered downstream discharges of water and sediment, a classic power differential between upstream and downstream riparian countries. In 2016, river levels fell so low that Vietnam was forced to beg China to release enough water from its dams to prevent severe crop losses on the Mekong River Delta.

China proposed a new model for regional governance of the Mekong River in 2014, at the seventeenth ASEAN-China Summit in Myanmar. Just sixteen months later the Lancang-Mekong Cooperation Framework was signed by the leaders of all six riparian countries (China, Myanmar, Laos, Thailand, Cambodia, and Vietnam), with China as the permanent chair.

It was the first regional intergovernmental organization that China has ever led in Southeast Asia. Its creation instantly raised speculation about the future of the MRC and brushed aside no fewer than three other competing regional cooperation frameworks seeking traction in Southeast Asia. These included the Lower Mekong Initiative, first proposed by the United States in 2009 and consisting of the United States and the same five riparian countries *except* China, and the 2009 Mekong-Japan Summit proposed by Japan, also with the same five countries.

Like the Mekong River Commission, the Lancang-Mekong Cooperation Framework is more a governing body than a treaty. It holds regular meetings at multiple levels of government, with heads of state required to assemble every two years and foreign ministers annually. While its nominal theme is "Shared River, Shared Future," its mandate is far broader. Stated goals include establishing cross-border cooperation in law enforcement, terrorism, tourism, poverty, agriculture, climate change, disaster response, and banking. China will provide more than $1.5 billion in loans and a $10 billion credit line to invest in regional infrastructure throughout Southeast Asia. This money will be used to build waterways, railroads, and highways linking

Southeast Asia with China. The aims of this new framework thus extend well beyond river management.

This sequence of events reveals how transboundary governance of the Mekong River is both an objective and a vehicle of power in Southeast Asia. In 1995, a growing need for a collaborative international process to manage competing visions for the river brought about the Mekong River Commission and two decades of relatively good cooperation in the region. Today, with the MRC's weakness exposed by Laos' hunger for hydropower, a new river-governance model is rising with a far broader mandate. A pressing need to equitably share the Mekong has thus become a strategic vehicle for something even bigger: a broader integration of Southeast Asia, with China as a leading backer of development in the region.

As political borders, rivers drown desperate people and modify the sizes and shapes of nations. As conduits of water, they stoke anxiety and power imbalances among neighbors. And in global affairs—from water-sharing pacts to cooperative governance to sweeping visions of regional economic integration—we see that rivers unite far more than they divide.

Except in war.

CHAPTER 3

THE CENTURY OF HUMILIATION AND OTHER WAR STORIES

From 2014 to 2019, a brutal new society rose and fell along the ancient civilizational cradle of the Tigris and Euphrates Rivers. The Islamic State, a violent militant jihadist organization variously known as ISIS, ISIL, or Daesh, emerged from the bedlam of the Arab Spring revolutions to briefly grow its dream of creating an ultra-orthodox Islamic caliphate in Syria and Iraq.

The heartland of this new theological civilization would emanate from a series of captured river towns and cities strung like pearls along the Tigris and Euphrates Rivers, together with their dams, hydropower, and surrounding oil wells and farmland. Ferocious even by jihadist standards, ISIS roared onto the global stage by capturing, in short order, the important Iraqi cities of Mosul, Qaim, Fallujah, and Tikrit, and even briefly threatened the capital, Baghdad. The group also capitalized on the disarray of Syria's ongoing civil war to consolidate power in the northwest, capturing much of the Euphrates River valley and the key Syrian cities of Deir ez-Zor, Al-Bukamal, and Raqqa, the latter of which ISIS declared the capital of its new caliphate. Nearly twenty thousand strong, ISIS fighters included not only Syrians and Iraqis but also foreigners from Saudi Arabia, Jordan, and Tunisia and as far away as Australia, France, Germany, the United Kingdom, and the

United States. The organization's zeal for beheading captured soldiers and Westerners and then posting videos of their executions online provoked global condemnation and fear.

At the peak of its power in late 2014, ISIS controlled some 10 million people and more than 100,000 square kilometers of territory in Syria and Iraq. It was aggressively raising cash from oil revenues, foreign donors, kidnapping ransoms, looting, and taxing its captured towns and cities. ISIS even threatened destruction of critical dams on the Euphrates to impose control upon its captured citizenry. Oil revenues alone were adding $1–2 million a day to its estimated $2 billion in assets. This money paid for weapons, vehicles, and fighters and funded social media propaganda campaigns that inspired dozens of overseas terrorist acts around the world.

In Orlando, Florida, a man declaring allegiance to ISIS entered a gay nightclub armed with an assault rifle and a Glock semiautomatic pistol, and calmly shot more than one hundred people in what was briefly the worst mass shooting in American history. In Nice, France, another sympathizer plowed a truck through a beachfront Bastille Day celebration, killing or injuring more than four hundred. A steady drumbeat of less-reported shootings, bombings, beheadings, deliberately crashed vehicles, and other atrocities piled up casualties in Afghanistan, Algeria, Australia, Bangladesh, Belgium, Bosnia and Herzegovina, Canada, Denmark, Egypt, France, Germany, Indonesia, Israel, Kuwait, Lebanon, Libya, Malaysia, Nigeria, Pakistan, the Palestinian Territories, Russia, Saudi Arabia, Tunisia, Turkey, the United Kingdom, the United States, and Yemen. By mid-2016 more than twelve hundred people *outside* of Iraq and Syria had been killed in terrorist attacks inspired or coordinated by ISIS.

The world reacted, and a U.S.-led coalition rained missiles onto the aspiring caliphate. By early 2018 nearly thirty thousand airstrikes had been carried out against ISIS targets in Iraq and

Syria. On the ground, the jihadists were gradually driven out by ground forces, with Russia-backed Syrian government armies advancing from the west and U.S.-backed Iraqi and Syrian opposition forces from the east. The war continued through both the Obama and Trump administrations, and in March 2019 the latter declared victory when U.S.-backed Kurdish forces ejected the last ISIS fighters from Bāghūz, a small Syrian border village on the banks of the Euphrates.

The ferocity and global reach of ISIS won it continuous, in-depth media reporting throughout this five-year period. This included sustained attention of professional cartographers from *The Economist* and *BBC News,* which published maps chronicling the expansion, oscillations, and contraction of ISIS-controlled territory as it was eroded by air and ground assaults. I took an early interest in the conflict and avidly followed these maps. As the war evolved, I was continually struck by how critical the region's two major waterways—the Tigris River in Iraq and Euphrates River in Syria—were to the organization's regional ambitions.

From the start, controlling these river corridors was clearly a key objective for ISIS. Geographically, the region's population centers and rich irrigated farmland hug the rivers. Hydropower dams supply most of its electricity, especially in Syria, where the power grid is less centralized than in Iraq. Culturally, these lowland valleys are dominated by conservative Sunnis, who were generally more tolerant of ISIS' strain of Salafist fundamentalism. To extract further cooperation from the populace, ISIS used river dams as weapons, both by withholding the water supply and by threatening to blow up the structures.

Map after map revealed the heart of ISIS power following the Tigris and Euphrates river valleys. The Euphrates, in particular, remained in ISIS control even after its territory had withered to 2 percent of its peak extent. After Raqqa fell and military operations ground toward a finale, the final maps showed the Islamic State reduced to little more than a long, snaking corridor following the river.

Cross That River

That rivers were important to ISIS jihadists is clear. But their war was far from the first to be shaped by these natural features. Since antiquity, other military conflicts have also been influenced by them. The preceding two chapters described how societies value rivers for natural capital, access, territory, and power. In this one, we find that these same qualities can make rivers strategic in times of war.

In a brutally direct application of wartime power, rivers have even been used as instruments of mass execution. In 1793, many tens of thousands of France's Vendée people, in the western Loire Valley, were slaughtered when the region's Catholic priests refused to support the new government of the French Republic during the French Revolution and Reign of Terror. The response from the new government was punishing and swift. In Nantes, a major city along the Loire River, the commander Jean-Baptiste Carrier ordered his soldiers to massacre as many Vendée men, women, and children as possible. With hideous efficiency the Loire River itself was conscripted in a campaign of mass drownings called the *Noyades des Nantes.* Civilians of all ages were stripped naked, bound, and ferried out into the river in barges to be sunk, or prodded underwater with bayonets until they drowned. An estimated five thousand people were massacred in the Loire River as part of this broader campaign of terror against the Vendée people.

Where rivers are used as political borders, they can become emblematic of bold leadership when warriors dare to cross them. Take, for example, a certain Roman conqueror and governor of

During the 1793–1794 Reign of Terror, France's new republic crushed a revolt among its Vendée people in the Loire River Valley. In the city of Nantes, the river itself was conscripted to execute thousands of royalist sympathizers, using specially modified barges from which bound victims could be efficiently drowned.

Gaul who decided to march his army south toward Rome in 49 BCE. This required crossing the Rubicon, a small river in what is today northern Italy, which marked a provincial political border. Crossing it with soldiers was expressly forbidden by Roman Republic law. To do so would be treasonous and an irreversible act of war.

Julius Caesar's decision to cross the river plunged the Roman Republic into a civil war, from which he eventually emerged victorious. Five years later he would be assassinated, but not before he had launched a series of far-reaching political reforms that would transform the Roman Republic into the vast Roman Empire. According to legend, Caesar uttered the words *"Alea iacta est"* ("The die is cast") as he waded into the Rubicon's forbidden waters. To this day, *Crossing the Rubicon* is a commonly used expression for describing a major decision from which there is no turning back.

Another example comes from one of the few widely recalled moments of the American Revolutionary War. Indeed, the United States might not exist today had George Washington not led a surprise military attack across the ice-choked Delaware River near present-day Trenton, New Jersey.

It was Christmas night in the year 1776. Washington's Continental Army was tattered and demoralized after losing numerous battles with British troops dispatched to quell the rebel uprising. New York City was captured, and American fighters had retreated deep into the Commonwealth of Pennsylvania. The Continental Congress, fearing capture and arrest of its members, had fled from its capital city of Philadelphia. British forces controlled New Jersey and had stationed a garrison of Hessian mercenaries at Trenton, a town across the Delaware River border.

Washington had lost most of his army and supplies. Defections were rampant and British victory seemed nigh. Sensing the war collapsing around him, Washington marshalled his remaining soldiers to launch a late-night surprise attack on Trenton, ferrying some 2,400 fighters across the Delaware River. A sleeting, hailing storm delayed their assault for a few hours, which turned out to benefit the attack because the mercenaries were sound asleep when the raid came from across the river at dawn.

Washington captured the garrison and proceeded to win another decisive battle at Princeton. His surprise victories reversed the momentum of the war, promoting the recruitment of new fighters into the Continental Army. Had the dangerous nighttime crossing of the Delaware failed, the American War of Independence would have likely been quelled, and the short-lived "United States of America" would be a brief footnote in history.

Washington's daring counteroffensive became enshrined in

By Christmas 1776, Britain was on the verge of crushing the American Revolution. Facing defeat, George Washington crossed the ice-choked Delaware River, the river border between Pennsylvania and New Jersey, in a do-or-die raid. The surprise attack captured Trenton and then Princeton, reviving the rebellion and reversing the course of the war. Decades later this painting, *Washington Crossing the Delaware,* created by the German artist Emanuel Leutze, would become an iconic image in American patriotic culture.

American lore and was later immortalized by a huge painting titled *Washington Crossing the Delaware,* which hangs today in New York's Metropolitan Museum of Art. It was created in 1850 by a German painter named Emanuel Leutze, who hoped, unsuccessfully, that it would inspire a pro-unification movement in the German Confederation. The painting was largely ignored in Europe but became an instant sensation in the United States, enjoying a star tour around the country. Just four months after it had settled into its permanent home in New York City, some fifty thousand people had paid good money to see it. Within a year, most American schoolbooks and households displayed some facsimile of *Washington Crossing the Delaware.* The larger-than-life painting of George Washington gazing fixedly at the enemy shore, his motley soldiers struggling to cross the icy river, became an instant and permanent icon of American patriotism.

America Divided

There once was a U.S. presidential election so divisive that when its outcome was settled, vast numbers of Americans felt alienated from their own country. Social divisions were raw and starkly geographical, with some parts of the country supporting the new president and other parts vowing to resist him. The root cause of the discord stemmed from deep-seated problems of economics and race in America, with conservatives seeking to preserve a traditional economy and way of life and liberals supporting progressivism and equality. A national choice between an old or new direction was at hand, and America was deeply polarized about which direction was best for the country.

This tension divided even supporters of the same political party, the Democratic Party, which fielded two strong candidates, each offering different visions for the future. Their bitter primary contest weakened both of them, helping an ungainly Republican to win the November election with less than half the popular vote. A feeling of deep unease hung over the country. Even before this new president took office on March 4, 1861, at least seven U.S. states had already aligned against him.

Five weeks later the United States was at war. The core dispute was an irreconcilable difference between slaveholder states and free states over whether slaves would be allowed in newly settled territories of America. Free states believed that the federal government could and should prohibit slavery in new territories. Slaveholder states disagreed, saying it was federal overreach and up to each territory to decide.

During his presidential campaign, Abraham Lincoln and his abolitionist Republican Party ran on a platform of banning slavery in new territories. After his election, the proslavery U.S. states of South Carolina, Mississippi, Florida, Alabama, Georgia, Louisiana, and Texas swiftly seceded to form a new nation called the Confederate States of America. Its capital was Richmond, Virginia, and Jefferson Davis, a recently resigned senator from Mississippi, was elected by acclamation as its first president.

Upon taking office, President Lincoln strained mightily to avert the coming war. In his inauguration speech on the steps of the U.S. Capitol, he assured slaveholder states that their institution was not in jeopardy. He promised to protect slavery in those states where it already existed. Most important, he implored, was preserving the federal union of the United States of America.

His plea was ignored. The newly seceded nation demanded control of all federal assets within their borders, including military bases. Lincoln refused. Fighting broke out on April 12, 1861, when Confederate forces opened fire on Fort Sumter, a federal military base in South Carolina's Charleston Harbor. Union troops had retreated there for safety, and their outgunned commander quickly surrendered. Within days, the state of Virginia seceded to join the Confederacy, and it was followed in short order by Arkansas, North Carolina, and Tennessee.

The bloodiest war in American history had begun. On July 21, the first major battle between Union and Confederate forces took place along Bull Run, a small Virginia river. In general, the Union would name battles after the rivers and creeks along which they occurred, whereas the Confederacy tended to use the names of nearby towns. For this reason, the first battle is called both the First Battle of Bull Run and the First Battle of Manassas.

America's civil war would last four very long years. More than 3.2 million soldiers fought in some ten thousand battles and skirmishes across the United States. They fought for land and waterways controlled today by twenty-three U.S. states and the District of Columbia, from North Dakota to Vermont to Florida, from the eastern seaboard to Texas and New Mexico.

When it was all over, the Union was still intact, 3.5 million slaves had been freed, and the president was assassinated. Some 620,000 soldiers were dead. That is roughly comparable to the cumulative death toll of every other war that America has ever fought, including the Revolutionary War, the Spanish-American War, the Mexican War, the War of 1812, World War I, World War II, the Korean War, the Vietnam War, and recent smaller wars in Iraq, Afghanistan, and Syria. To further appreciate the scale of this death toll, consider the fact that the United States had just 31.5 million people at the time, less than 10 percent of its population today. Nearly every town, every family, lost someone—if not more.

And it could have been even worse, were it not for the Mississippi River.

~

To the dismay of Northerners expecting a brief war, the Confederate armies were well led and highly effective against their better equipped neighbors, despite being outnumbered nearly two to one. Led by skilled generals like Robert E. Lee and Thomas "Stonewall" Jackson, they outmaneuvered their hapless Union counterparts, and a series of disastrous battle losses racked up casualties to levels unthinkable at the war's outset. Northern support for "Lincoln's War" ebbed, and his political opponents began calling for recognition of the Confederacy. In the 1862 midterm elections, the Republican Party suffered withering losses to Democratic challengers. Meanwhile, the dead piled high along fluctuating front lines in Virginia, Mississippi, and Tennessee. His political support fading, President Lincoln pressured his generals to do something—anything—to turn the war around. One of them, Ulysses S. Grant, understood that this required winning control of the Mississippi River.

The strategic importance of that vast watery artery—appreciated since the early days of La Salle, Washington, and Jefferson—was obvious to both sides. In 1861 rivers and rail were America's highways, and the Mississippi and its tributaries its superhighways, connecting the interior of North America to itself and to other countries. For Northerners, this access allowed commerce and export of industrial and agricultural products from the upper Midwest. For Southerners, the river brought food and goods from the northern states and carried off cheap plantation cotton for cash. Ships could pass from the continental interior to the populous East Coast and the rest of the world. The Mississippi River provides access, well-being, and strategic power to a magnitude that is largely overlooked in America today.

Territorially, the Mississippi River also cut right through the heart of the newly formed Confederate States of America. It separated its large western states of Texas, Louisiana, and Arkansas from its eight states to the east. As war grew imminent, slave state governors and the new Confederate president scrambled to install forts and gun batteries along its shores. The mightiest of these were positioned at Vicksburg, Mississippi, a citadel city perched on high bluffs overlooking the river, from which heavy cannons were aimed at the water with devastating precision and force. It was known as "the Gibraltar of the Confederacy." Together with Port Hudson, another fortified river town about 150 miles downstream, it guarded the most heavily defended stretch of the Mississippi River.

Vicksburg was a veritable fortress. It blocked Union ship access from the north while safeguarding free passage of fighters, weapons, and supplies within the Confederacy itself. So critical was Vicksburg that President Davis wrote to the commander of the army garrisoned there, General John C. Pemberton, with the following stern order: *"Vicksburg is the nail head that holds the South's two halves together! You must hold it at all costs!"*

Meanwhile, both presidents were testing a technological breakthrough in naval warfare. In March 1862, the world's first ironclad warships burst onto the world stage in the brackish confluence of the James, Nansemond, and Elizabeth Rivers in coastal Virginia. The two strange-looking vessels, the USS *Monitor* and the CSS *Virginia* (formerly the USS *Merrimack*), were heavily armed and sheathed in plate iron. They squared off and fired, one day after the *Virginia* had calmly destroyed two wooden-hulled Union ships, the USS *Cumberland* and USS *Congress.* They fought to a stalemate, each incurring only minor damage. A shockwave rippled through navies worldwide. The era of wooden-hulled warships ended, and the era of metal-hulled ships began.

Convinced of the significance of the new technology, the U.S. Navy contracted James B. Eads, a St. Louis civil engineer and businessman with experience building rivercraft, to build a special fleet of ironclad gunships designed for the Mississippi River. To reduce their draft, the vessels were built oddly broad relative to their length. Their sloping topsides were sheathed in metal plates to deflect bullets and artillery fire. Hatched doors along their sides could be raised to reveal bristling heavy guns to be fired at will from the water.

In early 1862 Union ironclads began projecting power down the Mississippi River and its tributaries. These strange gunboats helped capture Fort Henry and Fort Donelson, two Confederate fortifications on the Tennessee and Cumberland Rivers, and compelled the surrender of Memphis. They supported ground forces in the famous Battle of Shiloh. Then the vessels turned toward Vicksburg, with the aim of breaking the

Ironclad river gunboats helped the Union gain control of the Mississippi River and thus win the American Civil War. Shown here is the USS *Carondelet,* one of a fleet of river ironclads that helped to topple the riverfront city of Vicksburg, a crucial turning point in the outcome of the war.

Confederate blockade of the Mississippi and wresting control of the river all the way to the Gulf of Mexico.

Thus began a complex series of maneuvers and battles called the Vicksburg Campaign. The Union's fleet of river ironclads formed the backbone of the Mississippi River Squadron, led by the U.S. Navy admiral David Dixon Porter. The ground forces were commanded by General Ulysses S. Grant. On the Confederate side was Pemberton's army garrisoned at Vicksburg.

Through a series of feints and a great encircling movement, Grant looped west and south of Vicksburg, through the swamps of Louisiana. He then cut back east, bringing Union soldiers to an undefended location about forty miles south of Vicksburg on the right (west) bank of the Mississippi River (riverbanks are called left or right, facing downstream). In one of the most dramatic moves of the Civil War, Admiral Porter then raced his ironclads downriver past the blazing cannons of Vicksburg to rendezvous with Grant's forces, and then ferried them across the river to the left bank, onto Mississippi soil, in late April 1863. Grant's army then veered northeast and engaged in a series of fierce battles, capturing and torching Mississippi's capital city of Jackson on May 14, before swinging west again toward Vicksburg.

Pemberton's army was trapped. Cut off by Grant's infantry to the east and Porter's river flotilla, floating like fiery crocodiles to the west, he ordered his army to fall back into the city and dig in. Emboldened, Grant pressed for victory, driving his army against the city. He incurred terrible casualties. After two failures to break through Vicksburg's fortifications, he encircled it instead, laying siege while Porter's gunships shelled the city from the river. A long shipping supply line snaked down from the north, bringing food, ammunition, and clothing to the encamped Union forces.

Inside Vicksburg, the besieged soldiers and civilians began running out of food, water, and medicine. They dug caves to escape the constant shelling. They ate animal feed, horses, dogs, and rats. They began to starve and die.

On July 4, 1863—America's Independence Day—Pemberton surrendered his emaciated army. Grant promptly paroled them, and thirty thousand Confederate soldiers simply stacked their guns and wandered home. Five days later, Port Hudson was taken downriver. The Confederacy was severed in two, its east and west halves blockaded from each other. Union ships carrying fighters, guns, and supplies began steaming with impunity down the Mississippi superhighway, free to pass from Pittsburgh all the way to New Orleans and points east. The Gibraltar of the Confederacy had fallen.

Together with a major battlefield victory over General Robert E. Lee near Gettysburg, Pennsylvania, the fall of Vicksburg and loss of the Mississippi River spelled doom for the Confederate States of America. Four months later, President Lincoln chose the nearer Gettysburg site to deliver his Gettysburg Address, which would eventually become one of the most revered speeches in American history.

In it, Lincoln effectively redefined the purpose of the war. He recast it as a struggle for America's constitutional principles of equality, rather than as about the right to own humans as property. Seventeen months later he was assassinated, shot point-blank in the head, by the Confederate sympathizer John Wilkes Booth. His vice president and successor, Andrew Johnson, a Democrat, had a catastrophic presidency and lost the 1868 election to none other than a certain war hero, Ulysses S. Grant.

~

A forgotten footnote from the American Civil War concerns the USS *Saginaw*, a modest, shallow-draft paddlewheel gunboat. Launched in 1859 from the Mare Island shipyard near San Francisco, the *Saginaw* was the first U.S. Navy warship ever built on the West Coast. Despite being finished some two years before shots rang out at Fort Sumter, the ship was unavailable when war broke out. Instead of idling along the California coast or waiting in reserve at Mare Island, it was seven hundred miles up the Yangtze River, deep in China.

The Century of Humiliation

China is why the USS *Saginaw* was built. The reason this gunboat wasn't available when presidents Jefferson Davis and Abraham Lincoln went to war in the spring of 1861 is that it was already busy exchanging gunfire with Chinese batteries. As Ulysses S.

Grant plotted to break the Mississippi River blockade and end the American Civil War, it was steaming up the Yangtze into the heart of China. No one knew it yet, but the USS *Saginaw* was testing out a riverine military strategy that the United States, Britain, Germany, France, and other foreign powers would use to project military power over China for nearly a century.

This long story of the Yangtze River gunboats opens with Britain and its desire to illegally traffic drugs into China and forcibly open the country to international trade. In 1839, the world's greatest sea power attacked a supremely confident yet technologically outdated China. Britain was retaliating against the country's war on drugs. The cartels supplying these drugs were British merchants and their partners. They were selling massive quantities of opium—then smoked, today refined into heroin and opioid painkiller pills—in an illegal scheme to launder opium into silver, which they could then use to buy Chinese tea, porcelain, silks, and other goods for profitable resale back home in Europe.

At the time, China had the largest economy in the world. It possessed vast capacity to produce exotic goods desired in Europe, and huge domestic markets into which a rapidly industrializing Britain wished to pour its manufactured products. But to London's consternation, China remained firmly shut to foreign trade, and even visitors. Successive emperors of the Qing dynasty forbade outsiders from traveling or doing business in the country. A rare exception was Canton (now Guangzhou), a heavily regulated port at the mouth of the Pearl River (now called Zhujiang), in present-day Guangdong Province. A restricted, restive trade was tolerated there, but on the condition that only silver could be used to buy Chinese goods.

This arrangement worked for a while, but by the 1830s Britain was running short on silver and was awash in opium, grown cheaply in its India colonies. The medicinal use of opium had a long history in China, but recreational use of the drug was prohibited and rare. That changed when traders started shipping 150-pound chests of Indian opium to Lintin (now called Neilingding), a small island off Guangzhou. On Lintin the drugs were offloaded and redistributed to Chinese traffickers, who smuggled them onshore in small boats. From there, a network of drug dealers and corrupt officials ensured that the recreational smoking of opium spread rapidly throughout Guangzhou and into broader mainland China.

By 1837 a major drug crisis was underway in the country. Local law enforcement began intercepting and burning the smugglers' boats, but the epidemic still raged. China's emperor, Daoguang, alarmed by the explosion of addicts in his country, declared a war on drugs and sent a special commissioner named Lin Zexu to Guangzhou to shut down the flow of British opium into China.

As the chief warrior in this war, Lin Zexu was effective. He arrested drug dealers and corrupt officials and shut down the smoky opium dens. He confronted the foreign cartels and demanded that they surrender their drugs, which were piling up in warehouses due to the crackdown. When they refused, Lin Zexu confiscated more than twenty thousand chests of opium and dumped them in the sea. A diplomatic row ensued, and soon thereafter Britain set up a naval blockade across the Pearl River at Guangzhou.

So began the First Opium War, the first of two wars that began with British naval guns sinking Chinese junks at Guangzhou and ended with the forcible opening of China's markets and territorial sovereignty given to Western powers.

Right from the start, China's navy was technologically outclassed. British ships had steam power, rotating gun turrets, and explosive shells, a stunning advantage over the sails and fixed cannons of the Chinese navy. A particularly deadly vessel was the *Nemesis*, a shallow-draft iron warship commissioned in secret by the British East India Company. From its birthplace in a Liverpool shipyard the *Nemesis* slipped out to sea and headed straight to Guangzhou, where it exacted terrifying tributes among the channels and straits of the Pearl River Delta. London newspapers at the time showed the dark, low-slung vessel systematically exploding a succession of hapless war junks seemingly at will.

British gunships then took the war deep into China, steaming up the Yangtze to capture the important riverfront cities of Shanghai, Chinkiang (now called Zhenjiang), and Nanking (now Nanjing). Chinkiang was located at the Yangtze's confluence with the Grand Canal, an ancient, deeply important transportation waterway connecting the Yangtze with points north and Peking (now Beijing). By conquering these cities and controlling the Yangtze River, Britain slashed through China with its superior naval firepower.

China's forces were hopelessly outmatched. By 1842, Emperor Daoguang had little choice but to sign a punishing treaty at Nanking, the first of many lopsided treaties designed to force his country into the Western economic system. The Treaty of Nanking abolished the restrictive trade system that had existed at Guangzhou since 1760, turning it and four other cities—Shanghai, Xiamen, Fuzhou, and Ningbo—into free trade zones called "treaty ports." Chinese products had to be made available for export, and her domestic markets had to accept foreign imports. China would cede Hong Kong to Britain. She would pay heavy reparations for the war, including six million silver dollars to pay for the chests of opium that Lin Zexu had confiscated and destroyed. In return, China received next to nothing, and its opium-addiction crisis was ignored. Not to be outdone, the United States pressed China for similar trade concessions in a separate treaty signed in 1844. Unlike Britain, the U.S. treaty did permit the outlawing of opium sales but was otherwise similarly lopsided.

A peace of sorts descended for a few years, then deteriorated due to a combination of Chinese resistance to the treaties and Western hunger for even deeper concessions. On flimsy pretext, Great Britain launched a Second Opium War in 1857, again attacking Guangzhou and also Tientsin (now called Tianjin). This time, France joined in. The United States, while officially neutral, lent modest support to the European attacks.

The outcome was again militarily and diplomatically crushing for China. Britain forced a new set of treaty signings in Tientsin, requiring China to open even more treaty ports. Foreign ambassadors could work and live in Peking. Foreigners could freely travel and do missionary work inside the country. Through something called the most-favored-nation clause, other foreign powers were allowed to pursue similar concessions. In rapid succession France, the United States, and Russia all pressed China into signing similar treaties. More land was lost, including cession of 150 million square kilometers of northwestern and northeastern China to Russia, with the Amur River (Heilong Jiang) becoming the new political border between northeastern China and Russia (which it still is today). Again, China got next to nothing and its drug crisis was ignored. Indeed, just a few months after the treaties were signed, Britain pressured China into fully legalizing the opium trade.

Another two years of fighting would transpire before the Chinese government grudgingly accepted this new wave of treaties. British ships kept shelling Chinese forts until they did. Meanwhile, the Western powers set into motion a century-long project to force China to trade with the outside world. Still more treaties were pressed and signed, and by the early twentieth century, more than forty treaty ports were open for business in China.

Within these enclaves, foreign powers created municipal governments, businesses, schools, and courts. Foreign navies docked, businesspeople made deals, and missionaries traveled and proselytized inside the resentful country. Terrible rebellions broke out, then were quelled. The opium crisis raged on. What the Chinese today call the Century of Humiliation had begun.

Reading through the Treaty of Tientsin, two of its provisions leap out:

> British Ships of War, coming for no hostile purpose or being engaged in the pursuit of Pirates, shall be at liberty to visit all Ports within the Dominions of the Emperor of China, and shall receive every facility for the purchase of provisions,

and:

> British merchant ships shall have authority to trade upon the Great River [Yangtze].

It was an explosive situation, forcibly imposed on a proud country long accustomed to believing itself not only a dominant world power, but *the* supreme nation on Earth. And for many ordinary Chinese living outside of the treaty port cities, the most visible form of foreign occupation was the Yangtze River gunboats. Large and small, decrepit and modern, for ninety years from 1858 to 1949 an international flotilla of armed warships steamed along China's coasts and up the Yangtze deep into the heart of China.

For ninety years the foreign gunships came and went on the Yangtze. They were there to enforce the treaties and safeguard their enclaves. They were there to chase bandits, guard their nations' citizens, and protect their business and political interests. They were there to project diplomatic and military power over China.

At various times, gunboats of the British Royal Navy, the U.S. Navy, the French Navy, the Imperial German Navy, Italy's Regia Marina, and the Imperial Japanese Navy were patrolling these waterways. They had names like the *Yangtze Flotilla* (British Royal Navy) and the *Yangtze Patrol* (U.S. Navy). The treaties allowed these ships into China to enforce a globalized economic order. Gunboat commanders had the right to deploy "protective action" to safeguard their enclaves, and "punitive action" to avenge Chinese attacks on their citizens or businesses. As noted by the military historian Angus Konstam, "Gunboats were guarantors of Western commerce, privilege, and safety, and were often all that protected the foreign enclaves in China." They were a peacekeeping operation for protecting the stability of foreign interests and mercantilist trade.

The era of the Yangtze River gunboats came to an end when civil uprisings and a global war washed across China. After the collapse of the Qing dynasty in 1911, China's short-lived political unification under Sun Yat-sen and Chiang Kai-shek fell apart when Mao Tse-tung's Communists and Kai-shek's Nationalists polarized the country in a civil war. Their struggle, which Mao ultimately won (a story that also involves rivers, as we'll see in Chapter 4), was interrupted only by Japan's all-out invasion of China in 1937, which then escalated into the Pacific theater of World War II. In 1941, when Japan went to war with Britain and the United States, it was time for the Yangtze River gunboats to leave.

Their brief return after Japan's defeat was not tolerated by Mao's Communists, who were viscerally opposed to the old treaties. They fired upon Britain's gunboat HMS *Amethyst,* which barely escaped by running a gauntlet of Chinese artillery down the Yangtze, thus ending Western gunboat diplomacy in China.

Mao would go on to drive Chiang Kai-shek's Western-allied Nationalists out of the country, and by 1950 foreign access to China had shut down again, restricted to just Hong Kong, Macau, and the island of Taiwan.

In the United States, most people have heard of the Opium Wars, but few know much about them. This puts Americans at a disadvantage when trying to interpret the historical baggage and national aspirations of China, a country that will soon have the largest economy in the world once again. America's civil war lasted for four years and tore the country apart, leaving scars that still hurt. Even today, a trend of removing Confederate statues and memorials from public spaces is controversial. Bronze likenesses of long-dead generals must be hauled off secretly during the night to avoid protests and potential violence.

I wonder if America's rifts would be even deeper today had foreign gunboats patrolled the Mississippi River instead of Union ironclads, and had the Civil War lasted nearly a century rather than the duration of an on-time college degree. China's travails lasted from 1839 to 1949. Today, every Chinese schoolchild is taught about the *Century of Humiliation,* a multigenerational tragedy of foreigners, drugs, and river gunboats that utterly permeates the Chinese worldview and its dealings with Western power today.

It seems every war tale has a sidebar, and the story of the Yangtze River gunboats is no exception.

In 1941 the Empire of Japan launched a surprise air attack on the U.S. naval base at Pearl Harbor, Hawai'i, killing 2,403 Americans and wounding 1,178. Nineteen U.S. ships and more than three hundred planes were damaged or destroyed. The strike provoked America's formal entry into World War II with one of the last declarations of war actually made by the U.S. Congress. But

December 7, 1941, the date that President Roosevelt declared would "live in infamy," was not Japan's first attack on the U.S. Navy.

That distinction belongs to the USS *Panay,* a Yangtze River gunboat. The strike occurred four years before Pearl Harbor, during Japan's invasion of China and the chaotic run-up to the Nanking Massacre, in which Japanese soldiers slaughtered more than three hundred thousand Chinese. The *Panay,* with fifty-five sailors on board, was in Nanking to evacuate any remaining Americans in the city. The ship was anchored in the Yangtze River, showing American flags, when it was bombed and machine-gunned by Japanese fighter planes. Three crew members and civilians were killed and forty-eight wounded. The emperor of Japan issued an apology, claiming the attack was an accident, but survivors and historians insist that the *Panay* was clearly identifiable as a U.S. ship. The attack did not provoke an American retaliation, and today its sinking is largely forgotten.

Rivers of Metal

Between 1939 and 1945 the world was aflame in the greatest mass war of all time. It was a war fought in thousands of places, with nearly every country in the world somehow affected. It was a war in which an estimated 50 to 80 million people died. Details are still coming to light about this unfathomable conflict, including about rivers. Take, as one example, the strong possibility that the entire war might have been averted, if not for an act of bravery by a young German boy named Johann Kuehberger.

Kuehberger lived in the border town of Passau, across the Inn River from Austria. Children often played along the river, and one cold day in January 1894, young Kuehberger saw someone struggling in the water. Another boy had walked out on the river's thinly frozen ice, fallen through, and was drowning in the powerful current. Kuehberger dove into the opening and saved the other boy's life.

Word got around, and the rescue became something of a legend in town. Kuehberger went on to become a priest, and a colleague—another priest named Max Tremmel—described Kuehberger's recollection of the event shortly before his own death in 1980. However, the story remained uncorroborated until 2012, when a clipping from the local *Donauzeitung-Danube* newspaper was discovered in a German archive. While the article does not name the victim, close agreement between the newspaper clipping and Tremmel's testimony have led historians to conclude that the boy who narrowly survived drowning in the Inn River was very likely Adolf Hitler.

Fast-forward to 1939, the year Hitler opened World War II by invading Poland with 1.5 million soldiers, more than 2,000 tanks, and 1,300 aircraft. It was an overwhelming force against a poorly equipped Polish army possessing only a few dozen modern aircraft and armored vehicles.

One reason Hitler had so many airplanes was that Germany had ready access to a near-miraculous material that was disrupting the aviation industry. It was lightweight, flexible, and durable. Historically this material had been too energy-intensive to produce in large quantities, but as its value for building planes and other industrial applications became apparent, Germany made concerted investments in building hydropower dams and smelting facilities to become, by 1939, the world's leading producer of aluminum.

Aluminum airplanes of the German Luftwaffe soon controlled the skies over Europe. They began bombing armament production factories, power supplies, communications networks, railway yards, ports, canals, and other infrastructure. By 1940 it was painfully clear that air power was crucial to the outcome of World War II. Great Britain and the United States announced massive aircraft manufacturing programs, with the United States pledging to produce fifty thousand airplanes a year despite being officially neutral in the conflict. To achieve this, America's factories needed unheard-of quantities of aluminum. This, in turn, required access to bauxite, and huge quantities of cheap electricity for the smelting process.

Enter the Demerara River in Guyana (then colonial British Guiana) and the Saguenay River in Quebec. Alcan, a Canadian mining company, had access to large bauxite reserves in Guyana, and boats could be used to transport it down the Demerara River to a preprocessing plant at the riverbank town of Mackenzie (now part of Linden), for crushing and washing. From there, the washed ore could be transferred onto larger ships destined for Canada's St. Lawrence River seaway and its tributary, the Saguenay River, which drains Lac Saint-Jean and surrounding tributaries on the Canadian Shield. In 1941, in response to America's burgeoning demand for aluminum ingots, some deal-sweetening tax breaks from Ottawa, and political pressure, Alcan built a massive hydropower dam complex called the Shipshaw hydroelectric project on the Saguenay River.

Together with two other hydropower dams farther upstream, Alcan turned the Saguenay River valley into the world's leading producer of aircraft aluminum. Its aluminum production increased from just 75,200 metric tons in 1939 to more than 1.5 million metric tons by 1945, a twenty-fold increase in just six years. Bombers made from Canadian aluminum flew under Allied flags all around the world.

Aluminum was so vital to the Allied war strategy that it prompted the 1941 Hyde Park Declaration, a legal workaround concocted by Canadian prime minister Mackenzie King and U.S. president Franklin D. Roosevelt to allow the United

States—still officially neutral—to use raw materials from Canada to produce war supplies for Britain. Thanks to the complex at Shipshaw, Canada became a major wartime supplier of aluminum to the United States, and provided 90 percent of the aluminum used by Great Britain and its Commonwealth allies. In a global conflict fought with industrial capacity as well as troops, Saguenay River hydropower became one of Canada's most important contributions to the war.

Britain's Dambusters

Almost immediately after the war broke out, British Air Ministry planners began searching for ways to degrade Germany's industrial capacity. The Ruhr River valley, in particular, had extensive manufacturing and power-generation facilities that were the subject of intense focus in British bombing plans. A number of river dams supplied hydropower and water to this industrial heartland deep inside Germany. A plot was hatched to blow them up.

A key target was a large reservoir dam on the Möhne River. It was the most important supplier of electricity in the Ruhr Valley and held back the greatest volume of water. Two other targets were the Eder Dam, which generated electricity and maintained water levels in an important shipping canal, and the Sorpe Dam. Three others under consideration were the Ennepe, Lister, and Diemel dams. The British decided to blow up the first three in late spring 1943, when their reservoirs would be full, so as to achieve maximum destruction downstream.

Secret tests were conducted on an old dam in Britain's remote Elan Valley, east of Aberystwyth. Experiments quickly revealed that dropping even a massive bomb on top of a dam would not collapse it. Its upstream face had to be hit underwater, somewhere near the center. But the Ruhr Valley was too far inside

Germany to be approached on the ground, and the reservoirs were strung with steel nets to foil underwater mines or torpedo attacks. What was needed was some sort of airborne-launched device that could settle against the underwater upstream dam face, then trigger a tremendous blast.

After much testing, Royal Air Force engineers invented the "bouncing bomb," a 9,000-pound spinning cylinder designed to skip across the reservoir surface before sinking against the dam face. To enhance its bouncing ability, the bomb had to be back-spinning madly as it was dropped from a very low-flying aircraft. Like a spinning flat rock skipped across a pond, the device would hop across the water surface several times before slowing, sinking, and detonating.

A secret squadron of pilots and Avro Lancaster airplanes was assembled at Scampton Royal Air Force station, about seventy miles east of Leeds. The planes were heavily modified to carry one massive dam-busting bomb each, including a separate Ford V-8 motor engine to power the device's spin. The necessary low approach rendered barometric and radio altimeters useless, so the planes were fitted with two down-pointing spotlights angled in such a way that their beams would converge into a single circle of light on the water's surface when the plane was at the correct height to release the device.

The pilots practiced the low bombing runs for nearly two months, believing they were training to attack the *Tirpitz,* a huge German battleship. Only on the night of the raid did they learn that their orders were to fly low across the English Channel, the occupied Netherlands, and deep inside Germany to bomb the Sorpe, Möhne, and Eder dams.

The first Lancasters departed at 9:28 p.m. on May 16, 1943. They flew so close to the ground that one hit high-tension wires, erupted in flames, and crashed to the ground; its strange bomb would be retrieved and studied by German engineers the next day. Four out of five bombers heading to the Sorpe Dam were shot down or damaged. The one that did reach the target released its bouncing bomb but failed to breach the dam.

Eight of the nine bombers headed for the Möhne Dam reached their target a few minutes after midnight. As German defenders fired anti-aircraft guns, the first bouncing bomb skipped across the water and sank against the dam face as intended. A huge plume of water blew into the air but the dam was unharmed. The next bomb fell a few seconds too late because its plane had been shot and was on fire. The bomb skipped over the dam and exploded somewhere downstream. The third and fourth bombs blasted more water into the air to no effect. Moments after the fifth bomb blew water into the air, the dam face collapsed, releasing the contents of its nearly full reservoir onto the populated valley below.

Some 116 million cubic meters of water—roughly 50,000 Olympic-sized swimming pools—descended on factories and homes. They were demolished or submerged by the flood. The circling bombers reported seeing water raging down the valley and the lit headlights of cars growing dim beneath the deepening water.

Three remaining planes still had bombs and continued on to the Eder Dam. The first bounced twice and produced nothing more than a plume of water. The second struck the crest of the dam, exploding in a fiery flash that also damaged the bomber, which limped away and was shot down. The third bomb bounced three times, sank against the dam, and blew it apart. Another wall of water raged down another valley and onto its inhabitants below.

The breaching of the two Ruhr Valley dams killed 1,294 people and destroyed or damaged eleven factories and more than a thousand homes. Two power-generating stations were damaged, and bridges and buildings up to forty miles away were destroyed. The rivers became choked with sediment that had been trapped behind the dams, making them unnavigable. The manufacturing capacity of the German industrial heartland was damaged. Britain lost fifty-three airmen and eight Avro Lancaster bombers.

News of the Royal Air Force "Dambusters" made headlines around the world. Two days later, Prime Minister Winston Churchill praised the raid before a wildly cheering U.S. Congress. Britain continued to deploy its Dambusters throughout the war, attacking a German railroad tunnel in Normandy, E-boat installations, the Dortmund-Ems Canal, and even helping, at last, to sink the *Tirpitz*.

There simply is not enough space in this little chapter to recount all the ways that rivers influenced the strategies and battle tactics of World War II. The Volga River, for example, played a substantial role in one of the bloodiest battles in human history. Linking inland Russia with the Caspian Sea and Baku oilfields (in present-day Azerbaijan), the Volga was a central transportation corridor for the Soviet Union, which is why Hitler ordered his Wehrmacht 6th Army of some 200,000 troops to attack its significant riverfront city of Stalingrad (today called Volgograd), about 580 miles southeast of Moscow. Hitler was intent on seizing the Baku oilfields, and control of the lower Volga River would deny access to Soviet troops seeking to defend them. The city itself was also an important manufacturing and transportation hub and, being named after the Soviet leader Joseph Stalin, a tempting symbolic target for Hitler.

When the Wehrmacht attacked Stalingrad in August 1942, its Soviet defenders withdrew and then surrounded the elongate city, trapping the 6th Army against the mile-wide river. Hitler sent reinforcements, and a prolonged, brutal siege and mutual slaughter ensued as Russians and Germans fought street to street and floor to floor as snipers shot from the rooftops. Mamayev Kurgan, a stronghold hill position located above the city, was fought for so intensely that it changed hands more than a dozen times. Shellfire so churned the hillside that its soil filled with metal splinters and remained black throughout the winter, the snow melted away by explosions and fire.

Unable to escape and cut off from supplies, some quarter-million Axis soldiers were lost and perhaps four to eight times as many Soviets. Stalin forbade evacuation of Stalingrad's civilians, believing their presence would encourage the Red Army to defend the city more vigorously. The starving German survivors who ultimately surrendered a half year later were sent to gulag prison camps, where most perished. Only six thousand soldiers from the original Wehrmacht 6th Army ever returned home to Germany.

With combined Soviet and German casualties estimated at more than 1.5 million, Hitler's attempt to control the Volga River thus consumed one of his most powerful armies. The siege of Stalingrad, with the river itself helping to trap Hitler's forces, halted Germany's advance into the Soviet Union and was a turning point in the war.

Another example of how rivers influenced the strategies and battle tactics of World War II was Operation Market Garden, a dramatic September 1944 attempt by the Allies to capture a fifty-mile corridor of river and canal bridges and push across the Rhine River into Germany. It was the largest airborne assault ever launched, with 35,000 British and American paratroopers landing behind German lines to seize water crossings over the Waal, Dommel, Rhine, and other river and canal bridges. German defenders suffered heavy casualties but repelled the attack, thus prolonging the war in Europe. Other consequential World War II battles were fought for the Meuse, Dnieper, Narva, and Oder Rivers. The Battle of Sedan, fought for bridgeheads over the Meuse River, in particular, was extraordinary.

The Matador's Cloak

The lands along the Meuse River are among the most blood-soaked in all of Europe. From its source on the Langres Plateau near Pouilly, the Meuse winds north and east across France, Belgium, and the Netherlands before sinking into the North Sea. It is navigable along much of its length and, together with some radiating canals, is one of the more significant transportation waterways of Europe. But in wartime, the unusual physical geography of this river makes it especially strategic.

For several hundred kilometers, the river carves steep-sided bluffs into a great geological massif that rams into northeastern France, separating the rugged terrain of the Ardennes Forest from open flat plains running west to Paris and the rest of the country. This stretch of the Meuse, together with the Ardennes, has long posed a natural barrier to armies seeking to invade France from Germany and Eastern Europe, and been a traditional borderland separating past Germanic and Francophone empires.

France's reliance on the physical difficulty of descending the steep stone bluffs and then crossing the Meuse River led to the area's light fortification. For France's enemies, the high-risk / high-reward temptation of sending an army through this difficult terrain onto France's flatlands, with a straight shot to Paris, has resulted in some of the biggest military surprises and death tolls in European history. Over the past 150 years at least four major military conflicts have transpired along the Meuse valley and the rugged Ardennes massif northeast of the river.

German forces successfully crossed the Meuse in the 1870 Battle of Sedan during the Franco-Prussian war. Forty-six years later, the amassed armies of Germany and France fought to an agonizing German withdrawal in the 1916 Battle of Verdun, with a million soldiers dying over a postage stamp of ground. In 1944, American forces barely contained Nazi German troops in the Ardennes Forest, in a bloodbath that became known as the Battle of the Bulge. It was Hitler's last great counteroffensive, a failed attempt to repeat a devastating breach across the Meuse that had shocked the world four years earlier, and remains one of the most stunning maneuvers in military history to this day.

It was spring 1940, and all of Europe was readying for war. Germany's 1939 invasion of Poland had prompted declarations of war from Britain and France and triggered a Soviet invasion of eastern Poland. After Hitler and Stalin agreed to partition Poland, Stalin invaded Finland on November 30, 1939. Finnish soldiers on skis fought the Soviet forces along the Russia-Finland border throughout that winter, sometimes in minus-40-degree weather, in the so-called Winter War. In April 1940 Germany attacked the neutral countries of Denmark and Norway. The Allied forces of France, Britain, and exiled Polish troops readied for battle and the officially neutral United States began churning out weapons, airplanes, armored vehicles, and supplies to sell to them in anticipation of a major Allied attack on Germany sometime the following year.

At the time, France was Europe's greatest military power. The commander in chief of her army, Maurice Gamelin, believed that a German invasion would come from the north, through the so-called Low Countries of Belgium, Luxembourg, and the Netherlands (so named because they are located on ancient, low-lying delta deposits of the Rhine and Meuse Rivers). Like most landscapes that are deposited and reworked by rivers, the Low Countries had flat terrain and straight roads, perfect for a fast blitzkrieg invasion by Germany's motorized infantry and tanks. Also, the Low Countries had been Germany's preferred attack corridor during World War I, and spies confirmed that Hitler was amassing military forces near their borders.

Memories of World War I hung heavy in the air, and Gamelin was anxious to avoid the return of trench warfare on French soil. In anticipation of a German attack through the Low Countries, he readied his armies and equipment to send north. On May 10, 1940, German infantry rolled into Belgium and the Luftwaffe bombed Belgian fortifications and the city of Rotterdam. Four days later, the Netherlands surrendered, and Hitler's invasion of France through the Low Countries seemed imminent. Commander Gamelin executed his counterplans, sending still more troops, armor, and supplies north.

But it was all a ploy. As France and the Allies readied for a prolonged World War I–style trench war in the flat deltaic plains, long convoys of tanks and motorized infantry were snaking through the Ardennes Forest. A huge mechanized force was inching toward the most lightly defended stretch of France's border, aimed at Sedan and other bridgeheads across the Meuse River. In the vanguard of this army were numerous tank divisions, led by generals Heinz Guderian and Erwin Rommel.

When Commander Gamelin became aware of the German activity in the Ardennes Forest, he thought it was a feint. It could not possibly be Hitler's real target, he reasoned, because it would be too difficult to move tanks through the rugged massif and down the steep bluffs cut by the Meuse. He was disbelieving still, as the panzer divisions erupted from the forest and descended upon the river towns of Sedan, Monthermé, and Dinant from the east.

They were the spearhead of a major invasion force that would pour through the heart of France. Like the matador's cloak, the real feint was the faked invasion through the Low Countries, followed by a stabbing thrust through France's lightly guarded flank. The plan, later known as the *Sichelschnitt* ("sickle-cut"), was originally conceived by a Wehrmacht field marshal named Erich von Manstein.

After pounding the Meuse bridgeheads' limited defenses from the air, the panzer divisions lunged toward the river. On May 12, General Guderian's panzers attacked Sedan. Rommel took the crossing at Dinant. Terror and disorder fell upon the lightly manned French pillboxes, some of them inoperable because they had been locked shut by departed French garrisons ordered north to Belgium. Their defenses disintegrating, the French blew up what bridge crossings they could and withdrew.

The seemingly impassable natural barrier was breached, and Gamelin learned of his fatal error. Pontoon bridges were laid across the Meuse, and German tanks and motorized infantry began streaming across them into France. By May 16, Guderian and Rommel had raced their panzer divisions more than fifty miles inland, almost a third of the way to the English Channel. The amphetamine-chomping panzer crews were overextended and low on fuel, but the French were in total disarray and unable to retaliate. The country's best forces and equipment had been sent up north to support Belgium.

On May 15, France's prime minister Paul Reynaud called Winston Churchill, Britain's new prime minister, who had just taken office five days earlier, and told him that France was defeated. Churchill flew to Paris the next day and found government officials burning documents and preparing to evacuate the city.

But the panzers were not yet interested in Paris. After a couple of days for rest and repairs, they abruptly swung north, circling behind the Allied armies amassed in Belgium. The French, British, Belgian, and Dutch armies were trapped against the English Channel. Only through an epic marine rescue operation were a third of a million Allied soldiers evacuated from the beaches of Dunkirk, thus keeping the war alive. On June 22, with more than half the country occupied, France signed an armistice with Germany in the same railroad car and location where Germany had surrendered to France in World War I. Hitler personally accepted the surrender, seated in the same chair that had been occupied by France's Marshal Ferdinand Foch when he accepted Germany's surrender twenty-one years before.

France had fallen, just forty-one days after German tanks overcame the natural physiographic barrier of the Ardennes Forest and steep valley wall of the Meuse River. Britain stood alone. Four long years would pass before the Allied D-Day invasion of Normandy marked the beginning of the end of Germany's occupation of Western Europe.

A Milk Run in Vietnam

I first met Richard Lorman at his home, overlooking Hingham Bay in Massachusetts. I shook his hand and cracked a lame joke about his apparent continuing fondness for the water. I was excited and a bit nervous about the conversation I knew was coming. It was the first time I had ever asked a combat veteran to recount his personal war experiences, so I wasn't sure what to expect.

To my surprise, he had a miniature model, made to scale and exquisitely detailed, of one of the riverboats he had lived on. As he spoke, the model helped me to understand what he was saying about how the vessel worked and the events that transpired on it. It had once been a World War II troop-landing craft, the type used to ferry Allied troops across the English Channel on D-Day. A flotilla of these boats is featured dumping soldiers onto Omaha Beach in the dramatic opening moments of Steven Spielberg's *Saving Private Ryan*. I recognized the deep troop well and squared-off bow, with a drop-down landing gate on one end.

Unfamiliar to me were the numerous additions and modifications that had been made to the vessel. Heavy, shielded M-60 machine guns lined each side. A helicopter landing pad capped the troop well. Three cylindrical gun turrets squatted near the stern, each large enough to enclose a gunner manning a .50-caliber machine gun or grenade launcher. There were walls of sandbags stacked around the turrets—for extra protection against rocket-propelled grenades (RPGs), Lorman explained, which could easily penetrate the one-inch armor plating. The surplus WWII landing craft had been transformed into a weaponized, riverine armored troop carrier (ATC), one of many that patrolled a maze of rivers and canals on the Mekong River Delta in some of the grittiest fighting of the Vietnam War.

From 1965 to 1971 the U.S. Army, Navy, and Coast Guard operated hundreds of riverine watercraft on the rivers and canals of southern Vietnam. There were ATCs, river patrol boats, and assault boats. There were river minesweepers, salvage craft, and refueling boats. There were huge floating motherships with supply depots, barracks, mess halls, mechanic bays, hospitals, and docks, to which a dozen or more smaller boats could tie up. Some ATCs were converted into giant water cannons to blast away Viet Cong positions on the riverbanks; others were converted into flamethrowers, to set the positions on fire.

It's one thing to read about these technologies in a dry military report full of acronyms. It's quite another to learn about them from a combat veteran who experienced some of the worst horrors of the Vietnam War. During his one-year tour of duty, Lorman lived entirely on U.S. Mobile Riverine Force boats, serving eleven months as an ATC gunner and one month on a flamethrower. During that year, he estimates, he engaged in fifty serious firefights and was fired upon at least 150 different times.

For a sense of what it was like, Lorman recommended watching the early scenes of the 1979 movie *Apocalypse Now,* which he says rather accurately reflect his experience.

Warfare on the narrow rivers and canals of the Mekong Delta meant constant fear of ambush. Bullets and rockets would erupt from the dense foliage lining the banks, often from just a few yards away. Sometimes the attacks were a trick, an attempt to consume the riverboat's ammo before it reached a larger ambush downstream, or to direct friendly fire onto United States or South Vietnamese Army soldiers nearby. Booby-trapped banana bunches tempted from trees leaning over the water. Lorman constantly scanned the water for suspicious objects. Most were floating corpses, but some were mines. Other mines were concealed on the riverbed and detonated via hidden wires running from the water up into the foliage. Others were quietly attached to boats' hulls by enemy swimmers.

A lone Viet Cong soldier once crept onto Lorman's ATC while it was moored at a supposedly safe location, next to an open field secured by lounging South Vietnamese soldiers. He and his companions were relaxing on the helicopter platform with the landing gate down. Suddenly, a Viet Cong soldier leapt up onto the platform, opened fire from point-blank range, and ran off unscathed. The volley somehow missed Lorman but killed two of his compatriots who were playing cards.

He described the horror of the flamethrower boats, nicknamed "Zippos" after the cigarette lighter (see color plate). I learned that contrary to popular perception, their primary task was not just to defoliate the riverbanks, but to kill. During a firefight, the Zippo would ram straight into the riverbank and unload flaming arcs of burning, jellied gasoline onto the enemy. Lorman described a young Viet Cong soldier on fire, running and screaming out into the open. The heat detonated the grenades on his body, blowing him apart.

Lorman saw many American soldiers killed or wounded on the twisting waterways of the Mekong Delta. And as a gunner for a million-dollar, state-of-the-art weaponized vessel, he doubtless killed and wounded far more Vietnamese, although we did not talk about that. His boat, called T-152-6, was lucky, taking only one direct RPG attack over the course of his yearlong tour. Its luck ran out just two days before Lorman and his five surviving original crew members were scheduled to finish their service and go home.

Like a bad movie, all the superstitious rules were broken that day. It was Friday the 13th, June 1969. The crew was "short"—meaning about to go home—one of the worst bad-luck omens of all. But after 363 days on the rivers, their commanding officer had assured them that their tour of duty was over. They were docked at a mothership and in high spirits, inventorying their boat's equipment and cleaning, repainting, and restocking it for their replacements. "We were finished," said Lorman. "All we had to do was get the boat ready for the new crew coming in two days." Then, an officer appeared and ordered them on one last brief mission. It was no big deal, just "a milk run," the officer insisted.

The crew nearly mutinied, but in the end, after much heated back-and-forth, they followed the order. Lorman, petrified, tore out the spongy inside padding of an extra helmet and jammed its shell on top of his own. A soldier watching him from another boat threw over a second flak jacket and pair of pants, which he also donned over his own. ATC T-152-6 slipped off the mothership and motored upriver for its assignment: Delivering a fresh platoon of soldiers to an island in the Bến Tre River, one of the most dangerous places in the delta. At the time, the nature of their destination was kept from Lorman and his fellow soldiers—it wasn't until years later that he learned about the danger of where they were headed.

They picked up the platoon. Some thirty fully equipped soldiers packed into the troop well. With them was a "genuinely friendly" medic, and Lorman asked him what time it was. "Ten thirty," came the smiling answer, followed by the din of AK-47 automatic-weapons fire and rocket-propelled grenade explosions.

The soldiers leaped to their feet and returned fire into the leafy riverbank, about two boat-lengths away. "I recall that my childhood snowball fights were at similar distances," Lorman told me. Years later, he would learn that the friendly medic was instantly killed. Lorman was unaware of the man's death at the time because a bullet had passed clean through his own outer helmet, split his inner helmet, and lodged in his neck vertebrae. Shrapnel had torn through his intestines, and one of his legs was badly damaged. Lorman slumped to the deck and heard at least two RPGs explode inside the troop well, setting it afire. "I felt a great crushing feeling," he later recounted. "I was essentially paralyzed and couldn't move, breathe, or speak out.... I repeatedly went in and out of consciousness among other wounded, some on top of me.... I went through the intriguing out-of-body gently-floating-toward-a-warm-beckoning-light experience. It is eye-opening real, believe me."

In seconds, the attack was over. Survivors called for medevac support and did triage. Lorman was mistaken for dead and tossed into a pile with corpses above and below him. Somehow the mistake was discovered, and he was slid onto a stretcher and loaded into a loud, violently shaking helicopter. He recalls moving his charred right hand and finding the slippery left hand of another wounded soldier. The two strangers grasped each other's hand during the hallucinatory flight as they were carried off to their new lives of hospitals and rehabilitation.

Richard Lorman, a permanently disabled veteran, never married and has no children. He wears hearing aids due to "boats, explosions, bars, nightclubs, metal shop, and construction." A teetotaler during the war—he stayed on the boat during rare shore visits to a base—he does drink today. He spoke with sadness about the young age, malnourished state, and wretched equipment of his enemy. But I heard no complaints from Lorman. "I enlisted," he said flatly. "I wanted an adventure, and I got one." He joked about a recent medical scan that found a bullet in his calf, undiscovered for almost fifty years. He lives in a beautiful house he built for himself on the Massachusetts shore. A few years ago, he was able to find online the wounded soldier who had gripped his hand during the surreal medevac flight. Each man was glad to learn the other had survived. But asked his thoughts on the broader purpose of his war, he just shook his head. Insane. It was all just insane.

But there *was* a broader purpose. The job of ATC T-152-6, Lorman's apocalyptic houseboat for the most defining year of his life, was the same as hundreds of other vessels in a brown-water navy sent to control a critical area of South Vietnam, a now-vanished country that was the proxy for a much larger global struggle between two competing political ideologies.

~

After World War II, many colonial territories fought to shake free from their European masters. In Southeast Asia, the First Indochina War was a bloody rebellion that broke apart French Indochina after the 1954 Battle of Dien Bien Phu. The region was split into Cambodia, Laos, a U.S.-backed South Vietnam, and a Communist North Vietnam. The arbitrary partitioning of Vietnam at the 17th parallel—modeled after the 1945 Soviet-U.S. partitioning of Korea at the 38th parallel—was negotiated in Geneva as part of the armistice to end the war. It was supposed to be temporary, but the Truman, Eisenhower, and especially Kennedy administrations all supplied financing and military support to South Vietnam as part of America's overall Cold War strategy of "containment" of communism.

The North Vietnamese and their sympathizers in South

Vietnam, however, wanted to reunify Vietnam into a single Communist country, modeled after China and the Soviet Union. By 1964 a North-backed Viet Cong insurgency was threatening to topple South Vietnam. This worried U.S. president Lyndon B. Johnson, who pressured Congress into passing the Gulf of Tonkin Resolution, granting the president military authority in the region and a legal basis to wage unrestricted war in Vietnam. Johnson immediately began deploying tens of thousands of troops overseas and by 1968 more than half a million Americans were in Vietnam.

For North Vietnam, the war was all about ousting the foreign occupiers and reunifying the country. For the United States, it was about preserving non-Communist South Vietnam, an artificial state created in Geneva. The thinking was that allowing the two entities to reunite would enable the growth of communism and increase the influence of the Soviet Union and China. The American objective was to preserve South Vietnam, not to invade the North, which would surely have provoked China into entering the war.

The fight was therefore in South Vietnam, which meant that Viet Cong troops, weapons, and supplies had to flow down from North Vietnam. To win, the United States and South Vietnamese forces had to control these supply corridors. The most famous of these was known as the Ho Chi Minh Trail. But because Vietnam is both a coastal country and a river country, with few paved roads or railways at that time, the best way to sever the supply chains was through naval interdictions.

In 1965 the United States began "Operation Market Time," an eight-year blockade of the South China Sea using navy destroyers, ocean minesweepers, Swift boats, navy patrol gunboats, and U.S. Coast Guard cutters to intercept North Vietnamese vessels seeking to reach South Vietnam. Owing to the roughly north-south orientation of the coastline, any sea approaches from the north had to come from the east, making naval interceptions highly effective. When Operation Market Time began, some 70 percent of Viet Cong supplies came through the South China Sea. Within a year, this flow was reduced to just 10 percent.

That left overland trails, inland waterways, and the Mekong River Delta. North Vietnam sent vast quantities of supplies down the jungle paths of the Ho Chi Minh Trail through Laos and Cambodia, with branches splitting off into South Vietnam. The southernmost of these linked to the Mekong River through Cambodia, from which materials and fighters were distributed across the Mekong Delta using small junks, sampans, and barges throughout its complex network of rivers and canals. Its 700-plus-mile maze of waterways became a critical conduit for the transport of soldiers, weapons, and supplies to the southern part of the country.

The Mekong River Delta was also highly strategic in and of itself. It was (and still is) the most important rice-growing region in Vietnam. It contained half of South Vietnam's population and was strategically positioned near the capital city of Saigon (now Ho Chi Minh City). By 1965, the Viet Cong were choking off the main source of rice for Saigon. The delta's watery landscape could not be defended with motorized ground transport. While the U.S. Navy enjoyed overwhelming technological superiority on the open sea, the delta's maze of narrow rivers and canals made winning control of the delta extremely challenging.

Seeking to control this porous area, the United States and South Vietnamese forces launched a series of riverine campaigns with names like Operation Game Warden, Operation Coronado, and Operation Sealords. None could pacify the Mekong River Delta as effectively as Operation Market Time had shut down the South China Sea. A cat-and-mouse game ensued as American riverboats stopped and searched sampans to intercept Viet Cong supplies. The delta gradually became a war zone as ATCs and helicopters shuttled American and South Vietnamese Army forces around the area.

Despite these counteroffensives, Viet Cong forces continued to use these waterways effectively to position supplies and fighters, including ahead of North Vietnam's ferocious and unanticipated Tet Offensive in 1968. Amid mounting casualties, as well as student protests in the United States, this coordinated attack on Saigon and other southern locations cratered U.S. domestic political support for the war. By 1971 the last of the Mobile Riverine Force operations had been turned over to the South Vietnamese military, and the United States formally exited in January 1973. Two years later North Vietnamese forces overran Saigon. At a cost of at least 1.3 million Vietnamese and nearly 60,000 American lives, Vietnam was reunified as a Communist country.

~

While rivers have rarely, if ever, motivated wars, we have long conscripted them as silent combatants in our conflicts. During World War II, their natural capital of hydropower helped Canada produce aluminum and placed Germany's Ruhr Valley in the dual-spotlight crosshairs of the Royal Air Force Dambusters. As political borders and defensive barriers they are de facto targets for capture, sometimes triggering historical turning points such as Caesar's crossing of the Rubicon, Washington's crossing the Delaware, and Hitler's crossing the Meuse. Their value for military access was measured in the volumes of blood spilled for the Mississippi River at Vicksburg and the Volga at Stalingrad.

For nearly a century, foreign navies patrolled the Yangtze, quelling rebellions and projecting power deep into a resentful China. The physical impossibility of securing a maze of channels and canals caused four years of horrific guerrilla warfare on the Mekong River Delta, with far-reaching traumas for all countries and individuals involved. ISIS grew and then fell along the Euphrates River, clinging to its banks down to the last Syrian stronghold.

From antiquity to America's wars, from the Century of Humiliation to two global conflagrations, from mass drownings to Vietnam to jihad, rivers have long been important in time of war.

CHAPTER 4

RUIN AND RENEWAL

On August 26, 2017, Hurricane Harvey made landfall near Corpus Christi, Texas, then dawdled over the state for four agonizing days. At Nederland a total of 60.58 inches of rain fell, smashing the previous U.S. record of 52.00 inches, set in Hawai'i in 1950. Never before in America had a weather station recorded five feet of rain from a single storm.

Houston, the country's fourth-largest metropolis and home to 2.3 million people, received up to four feet of rain. Buffalo Bayou, a sluggish river flowing through its downtown, leaped from its banks and joined swelling tributaries to flood low-lying areas of the city. Freeway overpasses were submerged. Rescue boats motored down Houston's underwater streets as people climbed onto their rooftops to escape the rising waters.

Thirty miles to the northeast, the San Jacinto River swamped the tony neighborhoods of River Terrace and Northwood Country Estates. In Fort Bend County, the Brazos and San Bernard Rivers went into record floods, triggering evacuation orders for nearly 200,000 people. In other Texas counties, the Lower Neches River, Tres Palacios River, Colorado River, Oyster Creek, Trinity River, Sabine River, Big Cow Creek, and Guadalupe River all went into record or near-record floods. Thirty thousand water rescues were made.

When Harvey finally dissipated, some forty thousand people

had fled to public shelters in Texas and Louisiana. More than 300,000 buildings and half a million cars were damaged. At least sixty-eight people were dead from drowning or other causes directly related to the flooding, such as being crushed under collapsing structures. A third of a million homes were without power in the baking heat.

Three weeks later, I visited some of these flooded neighborhoods. Debris was everywhere. I saw sodden mountains of mattresses and ripped-out sheetrock and wall insulation. People were camping outside their ruined homes. Electricity had been restored, but the houses' interiors were dark and lifeless. A cat gnawed at something unidentifiable wrapped in plastic sheeting. Black mold crept up the outside walls of what had once been a typical middle-class Houston neighborhood.

It was just one of thousands of damaged neighborhoods undergoing a frantic demolition. All across southeastern Texas, property owners and volunteers were ripping out walls, insulation, and flooring from water-damaged buildings. They were racing against time, hoping to dry out the buildings' frames and electrical systems before mold and corrosion ruined them completely.

I was there at the invitation of Team Rubicon, one of the first disaster-relief volunteer organizations to arrive after the rivers retreated. Team Rubicon was founded in 2010 after two former U.S. Marines, Jake Wood and William McNulty, gathered supplies and volunteers to help the victims of a 7.8 magnitude earthquake that leveled Port-au-Prince, Haiti. Dismayed to find themselves among the first rescuers on the scene, they decided to create a new disaster-relief organization that would deploy faster than traditional aid organizations, which they perceived as overly slow and cautious. They named it Team Rubicon (yes, after Caesar's crossing; see Chapter 3) and sought out military veterans to serve as the organization's personnel and volunteers.

Alongside its core mission of disaster relief, reintegrating combat veterans into society is an important dimension of their work.

Team Rubicon's long Texas days began with a logistics briefing on a rooftop parking lot atop their temporary command center, a vacant warehouse near downtown Houston. Surrounding the amassed volunteers were dozens of rental cars with magnetic Team Rubicon placards slapped on the doors. Downstairs, in temporary offices, a youngish crowd was milling about. There were maps on the walls, scribbled plans on whiteboards, and air mattresses tucked between desks.

Team Rubicon is a big outfit, with millions of donation dollars and thousands of volunteers serving multiple disaster-relief operations at any given moment around the world. Around 70 percent of their volunteers are military veterans. I was shown around by Bob Pries, a busy fellow whose cell phone rang constantly—understandably, as he was managing hundreds of volunteers arriving and departing from Houston's two major airports every day.

In the days following Hurricane Harvey's landfall, more than thirteen hundred volunteers per day volunteered for Team Rubicon's relief and recovery effort in Texas. They launched boat rescues, chain-sawed fallen trees, and cleared debris. They shoveled mud, demolished or repaired dangerous structures, and gave financial advice. With so much devastation across such a large area, these volunteers, and others from many other organizations and church groups setting up operations in Houston, were often among the first helping hands to show up in the stricken neighborhoods.

A half year later, a grim reality had settled upon hundreds of thousands of Texans, many of whom could not afford to repair or rebuild. The Texas World Speedway, a racetrack, was filled with ruined cars awaiting damage-claims processing by insurance company underwriters. Team Rubicon volunteers were still

In 2017, Hurricane Harvey killed sixty-eight people and damaged some 300,000 buildings in southern Texas and Louisiana. Total damages are estimated at $125 billion, making it one of the costliest natural disasters in U.S. history. Shown here are volunteers for Team Rubicon, a disaster-relief organization run by military veterans, who were among the first to assist the flood victims. *(Laurence C. Smith)*

in Houston helping flood victims rebuild. The damage estimate for Harvey had risen to $125 billion, making Harvey the second-costliest storm in American history.

It was topped only by Hurricane Katrina, which in 2005 slammed into the Gulf Coast and New Orleans, causing at least 1,883 deaths and $161.3 billion in damage.

Katrina hit the New Orleans area at 6:10 a.m. on August 29, 2005. Two hours later, the Mississippi River overtopped some levees. Pumps were overwhelmed. Levees and floodwalls were breached or failed entirely. Because the city lies in a bowl-shaped depression that is largely lower than the river, some 80 percent of the city filled with water, in some places almost 10 feet deep.

Four months later I visited the city's Lower Ninth Ward, which had been submerged for weeks. The destruction was total. There were endless blocks of sagging, crazily leaning houses. Mudlines, marking the peak flood level, were near or at the rooftops. The only signs of life were tall, lush weeds, chirping birds, and the hungry stares of feral dogs. Many homes were too waterlogged to save, and most have still not been rebuilt. More than half the levees protecting greater New Orleans were breached, damaged, or destroyed and some 95,000 residences were lost.

Along the Gulf Coast shoreline the storm surge washed up and over the beaches, plucking houses clean off their foundations and tossing cars like children's toys. Near Biloxi, Mississippi, I saw big-box stores blasted clean down to their bare, buckled steel framing. Crushed SUVs were hurled into swimming pools. Restaurants and homes had simply vanished, leaving behind only clean white slabs of concrete foundation. More than 200,000 buildings were destroyed.

Most people think of floods as extremely rare, unforeseeable "acts of God." This is not the case. Floods are a recurrent, expected, and anticipable occurrence all over the world.

The damages wreaked by Harvey and Katrina were simply extreme examples of a common phenomenon. In 2017 alone, the United States suffered no less than sixteen separate "megadisasters" (defined as incurring damages of $1 billion or more), including ten other floods and storms besides hurricanes Harvey ($125 billion), Maria ($90 billion), and Irma ($50 billion). Irma made landfall in Florida just days after Harvey struck Texas. Maria then laid down a swath of destruction across Puerto Rico and several other Caribbean islands in September, killing at least sixty-five people directly and nearly three thousand in the aftermath. Since 1980, the United States has suffered some 250 megadisasters, with damages totaling more than $1.7 trillion (after adjusting for inflation).

And those are just the big ones. Every year, countless smaller floods along river and creek bottomlands are triggered by snowmelt, rainy periods, and intense thunderstorms. They cause billions of dollars in damage and threaten life and property in every state in America and virtually every country on Earth. Globally, floods are deadly and disruptive, averaging more than five thousand deaths and more than $50 billion in damages every year.

Clearly, this takes a toll on human well-being. As we will learn in Chapter 9, nearly two-thirds of all humans alive today live near large rivers, so floods are a chronic danger and destroy lives and property with depressing regularity. Yet floods also provide natural capital by infusing river floodplains with fresh deposits of silt, nutrients, and water, creating some of the world's richest ecosystems and finest agricultural land. Where insurance or governmental disaster-relief programs are in place, floods trigger inflows of cash that spur economic growth and shape regional demographics. On rare occasions, upheavals caused by river floods can even unseat political powers or change legal norms. The rest of this chapter explores how far-reaching some of these unusual cases can be.

After the Deluge

After the Mississippi's levees broke, some 400,000 people were displaced by Katrina. An estimated 100,000 to 150,000 refugees moved to Houston alone within the first few months, spiking rents and abruptly raising the city's population by more than 3 percent. Some never returned, and twelve years later, many of these same refugees were clobbered again by Harvey.

Meanwhile, the city of New Orleans was stripped of people. Two months before the Mississippi poured into the city, the total population of Orleans Parish (containing the city of New Orleans) was 454,085. Five years later, its population was down nearly a quarter, to 343,829.

The racial makeup of this exodus differed from that of Orleans Parish as a whole. Home losses were worst in low-elevation, low-income neighborhoods—as exemplified by the Lower Ninth Ward, which was poor and black. Therefore, a disproportionately large share of the out-migration was poor and black. Most homeowners in these neighborhoods could not afford pricy flood-insurance premiums for their high-risk properties. People lost everything and, with no insurance money to rebuild, had little choice but to walk away.

The exodus was also disproportionately young. Young people are more likely to rent than own and young homeowners typically have fewer savings with which to rebuild. The sudden housing shortage—further exacerbated by an influx of disaster-recovery and construction industry workers who also needed housing—caused rents for the city's remaining rental stock to skyrocket. A rebuilding boom in wealthier neighborhoods attracted thousands of construction-trade workers of Mexican and Central American descent, and some stayed after the building boom was over. Five years after the flood, New Orleans found itself smaller, whiter, older, richer, and more Hispanic than before.

According to decadal U.S. Census data, the total population of Orleans Parish in 2000 was 484,668, of which 325,942 (67.3 percent) were black. Ten years later, the population had dropped to 343,829, of which 206,871 (60.2 percent) were black. Put another way, 85 percent of the overall population decline was black. While not all of this population loss can be attributed to Katrina, it is clear that New Orleans, historically a proudly African American city, became less black in part due to the city's flooding and subsequent housing crisis.

The population drop, however, was temporary. Today, for the

first time in a century, New Orleans is growing again. By 2017, the estimated population of Orleans Parish was up to 393,292, nearly 15 percent higher than in 2010. Incomes and employment had recovered, and rents had fallen back to normal. This rebound is entirely consistent with nationwide studies showing that floods (and other natural disasters) commonly *grow* populations, rather than shrink them.

Through a nationwide analysis of migration data, James Elliott, a professor of sociology at Rice University, found that while natural disasters definitely push marginalized people out, they also pull other marginalized people in. Asians and Hispanics, in particular, often move in large numbers to areas struck by natural disasters—even relatively small ones (defined as $51 million or less in damage). All across America, these traumatic events seem to trigger localized economic and population growth spurts. The greater the damage, the bigger the spurt, often even surpassing the pre-disaster status quo.

Cash inflows of outside recovery money, together with a disrupted local workforce and social fabric, create new employment opportunities attractive to newcomers having a hard time elsewhere. Insurance company payouts, federal disaster-relief loans, and charitable donation monies pour in. City planners dust off long-unfunded redevelopment plans to turn catastrophe into opportunity. New jobs materialize, not only for the construction trades but for planners, designers, engineers, and food services. Within a few short years, these transfusions of money and people can shape a flooded city's economy and demography in lasting ways.

Another phenomenon associated with river floods and other property-destroying natural disasters is a spike in criminal corruption. The 1993 "Great Flood" of the Missouri and Mississippi Rivers was one of the most destructive natural disasters in U.S.

history, severely damaging parts of Illinois, Iowa, Kansas, Missouri, Nebraska, North Dakota, South Dakota, Minnesota, and Wisconsin. Nearly $1.2 billion in U.S. Federal Emergency Management Agency (FEMA) recovery money (approximately $2.1 billion in consumer-price-index inflation-adjusted 2019 dollars) poured into these nine states to aid their reconstruction and recovery. This influx of federal disaster-relief money roughly tripled the number of corruption convictions for several years.

The reasons for this are straightforward. The aftermath of natural disasters is chaotic and urgent, creating ample opportunities for graft. To expedite recovery, standard requirements for competitive bidding might be suspended, for example. Typical disaster-zone corruption crimes include soliciting bribes for federally funded reconstruction projects, money laundering, kickbacks, cronyism, and embezzlement. The bigger the damage and cash influx, the bigger the spurt in criminal convictions. This observed correlation between recovery money and corruption has been documented nationwide, and raises the interesting possibility that historically high corruption levels in states like Louisiana and Mississippi might in part be due to the frequency and severity of flood disasters in these areas.

It's easy to assume that river floods do not discriminate. After all, nature knows nothing about class or race. But along low-lying river valleys and coastal deltas, flood-prone land is usually low-income land. After the floodwaters recede, poor people are often the most affected and have the least means to participate in the recovery. As the better-off file insurance claims and hire architects to rebuild, the worse-off depart even as new demographic groups move in. When river floods destroy buildings, they thus have a role alongside other social and economic factors in shaping the size and diversity of American communities.

When the Levees Break

Sometimes, floods have political consequences.

FEMA's agonizingly slow rollout of Hurricane Katrina rescue and recovery operations, for example, caused lasting damage to public perceptions of George W. Bush, the Republican president in office at the time. As ten thousand predominantly black flood victims sweltered in the New Orleans Superdome without food, water, or working toilets, Bush praised the beleaguered FEMA director, Michael Brown, for doing "a heck of a job." In fact, Brown's incompetence led to his resignation ten days later, and Bush was ridiculed as being out of touch and uncaring about black people—a perception that would dog his presidency and that bothered him greatly. In a Gallup poll conducted afterward, 60 percent of all respondents and a whopping 80 percent of black people responded "does not" to the question "Do you think George W. Bush does—or does not—care about black people?" Despite appointing more black Americans to high-ranking cabinet positions than any previous administration, the botched federal response to the flooding of New Orleans imparted lasting damage to the Bush presidency among minorities.

Bush was not the first president of the United States to suffer political fallout from the rising waters of an epic storm and the associated Mississippi River flood. The transformative yet strangely forgotten Mississippi flood of 1927 transfixed the country and helped elect a U.S. president. It cracked the first of a series of fissures that separated black Americans from the Republican Party, permanently changing the face of U.S. politics to this day.

Today, black Americans overwhelmingly support the Democratic Party. In the 2016 presidential election, Republican Donald J. Trump won just 8 percent of the black vote. Four years earlier, the previous Republican nominee, Mitt Romney, garnered a measly 6 percent against Barack Obama. Yet a century ago, the numbers were reversed—in favor of Republicans.

Frederick Douglass, a prominent black abolitionist and statesman, was a Republican. So was Abraham Lincoln, who as president issued the Emancipation Proclamation, led the country through a crushing civil war over the issue of slavery, and rammed through the Thirteenth Amendment of the U.S. Constitution to permanently abolish slavery in America. It was congressional Republicans who pushed through the Fourteenth Amendment granting former slaves U.S. citizenship and equal protection under the law, and the Fifteenth Amendment granting black men the right to vote. Fifty more years would pass before the Nineteenth Amendment granted women this same right—again championed by Republicans.

Democrats, in contrast, battled these amendments to the Constitution. They fought them in northern states, where they feared adding blacks to the voter rolls would help Republicans win elections. They circumvented them in southern states, by segregating blacks with odious Jim Crow laws. Blacks were unwelcome in the Democratic Party, and black delegates were not even allowed to officially attend the party's national conventions until 1936, more than sixty years after the Fifteenth Amendment entered force.

What happened? How did the Republican Party lose their overwhelming support from African Americans, which was as lopsided for them in the early twentieth century as it is today for the Democratic Party?

Most history books trace this story back to Franklin D. Roosevelt, the first of four progressive Democratic presidents. His New Deal strengthened social safety nets for Americans made destitute by the Great Depression, a demographic that included many blacks. Roosevelt was reelected with the support of 71 percent of black voters in 1936, although roughly half of them considered themselves Republicans. In 1948, his successor, Harry Truman, desegregated the armed services and outlawed racist hiring policies for federal jobs. This damaged his standing with white southern Democrats but won the support of 77 percent of black voters in his 1948 reelection bid, helping him eke out an underdog victory against his Republican challenger, Thomas Dewey. In 1963, the Democratic president John F. Kennedy was assassinated while pursuing new legislation to abolish segregation and Jim Crow laws. His successor, Lyndon B. Johnson, carried on with this effort and signed the Civil Rights Act into law on July 2, 1964. Four months later, Johnson was reelected in a landslide, winning 94 percent of the black vote. The transfer of black voting power from the Republican Party to the Democratic Party was complete.

Missing from this important history is the fact that African American defection from the party of Lincoln had begun even before FDR was elected. The origins of this disaffection trace to the massive 1927 flood of the Mississippi River, which upended the lives of hundreds of thousands of black farmworkers.

To appreciate the scale of this flood, it's important to understand that the Mississippi is a true leviathan. Its watershed, one of the largest on Earth, drains water from thirty-one U.S. states and two Canadian provinces. With an area of 1.2 million square miles, it reaches from Canada to the Gulf Coast and from Virginia to Montana. In total, it covers more than 40 percent of the conterminous United States.

Preconditions for a catastrophe were set throughout August 1926, when a series of storms pounded the upper Midwest, destroying the fall harvest and saturating the soil. More storms soaked the basin that autumn, winter, and spring, raising river levels from Illinois to Louisiana to record highs. In January 1927, the Mississippi's great eastern arm, the Ohio River, flooded its valley from Pittsburgh to Cincinnati. Two smaller tributaries, the Little Red and White Rivers, breached their levees and inundated the surrounding Arkansas farms with up to 15 feet of water. In March, a succession of flood waves coursed down the Mississippi River, driving thousands of people to the levees in a frantic effort to raise them with stacked bags of sand and earth. One flood after the other tested, saturated, and eroded the levees.

Then, on April 15, 1927—Good Friday—things got really nasty. A broad storm dumped between 6 and 15 inches of rain across several hundred thousand square miles of the Mississippi basin. With the soil now fully saturated, and streams and wetlands already swollen, virtually all of this new rainfall rushed straight into the branching veins of the Mississippi. In a sweeping historiography of the disaster, titled *Rising Tide: The Great Mississippi Flood of 1927 and How It Changed America,* John M. Barry writes the following:

> The river seemed the most powerful thing in the world. Down from the Rocky Mountains of Colorado this water had come, down from Alberta and Saskatchewan in Canada, down from the Allegheny Mountains in New York and Pennsylvania, down from the forests of Montana and the iron ranges of Minnesota and the plains of Illinois. From the breadth of the continent down had come all the water that fell upon the earth...down as if poured through a funnel, down into this immense writhing snake of a river, this Mississippi.

The flood crumbled the levees like brown sugar. It swamped the vast agricultural plains of the Mississippi River valley from Illinois to the Gulf of Mexico. More than 700,000 people were left homeless. An official death toll verified 313 drownings, but the true number was far higher. Many victims were swept out into the Gulf of Mexico or buried under deep deposits of sand and mud. If not for the Atchafalaya River, which siphoned off more than half the flood's discharge, the monster would have wiped out New Orleans.

The 1927 Mississippi River flood and its aftermath dominated the headlines of national newspapers for the rest of the year. Inexplicably, Calvin Coolidge, the Republican president, refused to visit the scene of the disaster. He curtly declined repeated and increasingly desperate entreaties from leaders in the devastated states and from relief organizations like the Red Cross, who knew that a visit from the president would elicit badly needed donations and volunteers from around the country and the world. It proved a fatal political blunder that has been scrupulously avoided by U.S. presidents ever since. Today, wherever a big natural disaster strikes, the president soon appears. Press conferences are held, and images disseminated of the president somberly conferring with relief agency leaders, praising first responders, and embracing victims.

Into this gaping political vacuum strode Herbert Hoover, Coolidge's obscure secretary of commerce. Hoover, a former mining engineer, took a keen and personal interest in the flood. He worked tirelessly and effectively as head of the nation's relief and recovery effort. He frequently visited the disaster zone, where he deliberately and effectively cultivated an extraordinary amount of media coverage. He insisted that newspaper photographers and reporters be given full cooperation and access to the ongoing recovery operations and to him. Within months, due to the immense scale of the disaster and his high visibility as the face of the nation's response, Herbert Hoover became a household name across America.

Nineteen twenty-eight was a presidential election year. To the dismay of establishment Republicans, Hoover's newfound fame catapulted him over his primary rivals, including the party

The largely forgotten 1927 Mississippi River flood ranks among the worst natural disasters in American history. It also had important political reverberations, helping to elect Herbert Hoover as the thirty-first president and creating the first cracks of discord between black voters and the Republican Party.

favorite, Frank Lowden, a former governor of Illinois. Herbert Hoover clinched the Republican nomination and went on to win the general election.

But not everyone was thrilled with Hoover.

All who lived in the flood's path suffered. But in a depressing precursor of what would reoccur seventy-eight years later during Katrina, the black and the poor bore the worst of the harm when the Mississippi burst its levees.

Slavery had ended, but black servitude to white landowners had not. With some $3 billion in human "property" erased from their balance sheets, the old wealthy class of slave-owner plantation masters was gone. In their place was a system of indebted white landowners utterly reliant on the labor of black sharecroppers. By keeping crop-shares meager and reselling food staples and supplies on credit to the sharecroppers, the landowners effectively chained their labor to the land. While technically free to leave, few sharecroppers had the means to do so. But the first wave of the Great Migration, in which some six million southern blacks ultimately left the South, had already begun. Many poor southern blacks, and especially sharecroppers on the Mississippi Delta, had a relative or a dream in some growing northern city like Detroit, Pittsburgh, or Chicago.

Into this changing time roared the 1927 flood, driving whites and blacks alike onto the levee tops and into refugee camps on high ground. In Greenville, Mississippi, a huge levee standing 8 feet high and 50 feet wide was unable to protect the town. Foaming waves tore through it, smashing against houses like combers on a beach before sinking the town under standing water. Thousands of survivors floundered up onto what remained of the levee crest, its long narrow strip the last high ground in a sea of muddy water.

The Red Cross eventually set up 154 refugee camps, called "concentration camps," in Arkansas, Illinois, Kentucky, Louisiana, Mississippi, Missouri, and Tennessee. Most were segregated, and all through that horrible summer black Americans were pressed into flood recovery labor. They filled sandbags and repaired levees. They unloaded Red Cross food rations for the white camps and their own. They distributed emergency supplies from relief barges sent down the river. In most of the camps, conditions were difficult but bearable. People received food, and, if they worked, one or two dollars a day.

That did not happen in Greenville, where black refugees were forbidden to leave, forced to work, and guarded by armed white men. When rescue steamers arrived to pull survivors off the levee, the town's white leaders forbade blacks from evacuating, to the captains' great surprise. The townsmen knew that if the laborers left, they would never return—something they feared even more than the flood itself.

In Greenville, some thirteen thousand black refugees were squeezed into a concentration camp perched on the levee. The camp was patrolled by armed guards. Desirable food supplies, such as canned peaches, were commandeered for whites. Blacks were forced to work without pay and wear tags on their shirts identifying their job assignments. They were forced to clean, cook, and do laundry. Refusal to wear a tag was punished by withholding food. Armed guards refused to release blacks from the camp until a landowner, his fields finally drained, came to retrieve them.

Many refused to be retrieved, believing the disaster had created the opportunity to leave. There were atrocities and beatings throughout that year's epic power struggle between white landowners desperate to retain their labor and a fed-up underclass of black sharecroppers. The 1927 flood thus accelerated the Great Migration with a massive exodus of displaced flood victims who knew the time had come to leave the South for good.

Through all of this, Herbert Hoover played an artful game. He made an outward show of sympathy for black Americans displaced and subjugated by the flood but took little action to resolve their plight. With an eye to the presidential nomination, he moved to consolidate the traditionally overwhelming support of black voters enjoyed by his Republican Party. As word got out about the abuses in Greenville, he invited Robert Moton, a prominent black leader and protégé of Booker T. Washington, to form an advisory panel to investigate them. In private, he dangled at Moton a plan for a vast "land resettlement" mortgage program, seeded with $4.5 million in Red Cross funding, that would grant 20-acre farmsteads to thousands of displaced black sharecroppers across the destruction zone. He would do something for black Americans "more significant than anything which has happened since Emancipation."

But President Hoover was lying. He knew there would be no land resettlement program and, indeed, refused to support it later. His only interest in Greenville was containing political damage from the scandalous atrocities that were happening there. He was playing Moton, and the black electorate more generally. They noticed.

In his 1928 election victory, he lost 15 percent of the black vote, a stunning first for the Republican Party. Hoover went on to nominate a Supreme Court justice so racist that his own party balked. When he ran for reelection in 1932, Robert Moton, one of his most steadfast black supporters, refused to endorse him. Hoover lost in a landslide to the Democratic nominee, Franklin D. Roosevelt. The first cracks had appeared in a political levee once thought to be indestructible. It was the beginning of the end of black Republicanism in America.

Weaponizing China's Sorrow

Several years ago, I was invited to give a keynote lecture in Taiwan. The venue was a cavernous auditorium in a historic, architecturally striking building named Zhongshan Hall in downtown Taipei. The structure is famous for two reasons. First, it is where Japan formally returned Taiwan to the Republic of China on October 25, 1945, after a bloody eight-year attempt to invade and conquer China. A month earlier, Japan had similarly surrendered to the Allied Powers on the deck of the USS *Missouri* in Tokyo Bay, to formally end World War II.

The second reason Zhongshan Hall is famous is for its sweeping open-air balcony, from which Chiang Kai-shek (Jiang Jieshi), the last non-Communist president of China, came back onto the world stage after being exiled from his country. In 1949, Chiang and his Nationalist Party (Kuomintang) were overthrown by Communist armies led by Mao Tse-tung (Mao Zedong), and Chiang fled to Taiwan with the remnants of his forces and government. Retaining the name "Republic of China," he turned the little island into a base of operations from which he plotted to recapture the Chinese mainland. From Zhongshan Hall's soaring outdoor balcony he delivered many rousing speeches and, later, inauguration addresses, as the leader of Taiwan.

The complicated stories of Japan's invasion of China, the exile of Chiang Kai-shek, and the rise of Communist China intersect on a major river in northern China, more than one thousand kilometers northwest of Taipei. In 1938, Chiang Kai-shek did something terrible to the great Huang He, or Yellow River, which helped bring about his own downfall and forever changed the course of Communist power in mainland China.

As described in Chapter 1, the Yellow River is both the cradle of Chinese civilization and the most dangerous river in the world. Through a quirk of geology, it sweeps through the great Loess Plateau, a massive blanket of windblown silt covering more than 600,000 square kilometers in north-central China. Soft and crumbly, the Loess Plateau is rapidly eroding, washing enormous volumes of silt into the river. This silt transmutes the water into a gritty, tan-colored liquid, giving the Yellow River its name and the highest sediment load of any major river in the world. It is a true freak of nature, carrying more than a billion tons of sediment to the ocean each year. That is roughly comparable to the annual sediment load of the Amazon, the world's largest river, despite having less than *1 percent* of the Amazon's discharge.

It is precisely this enormous sediment load that made the river so historically beneficial and dangerous to humanity. Not all of this silt washed out to sea. For millennia it was naturally spread across the landscape by floods and avulsions, creating some of the most fertile land on Earth and a natural epicenter for an agricultural civilization to develop on the North China Plain (recall from Chapter 1 the legend of Yu the Great, whose mastery of one such avulsion triggered a cultural flourishing and the First Dynasty of China).

Floods destroy as well as build, of course, so farmers and eventually governments built levees to control the Yellow River. Their objective was to protect the surrounding villages and pin the shifting river in its course. But sediment continued to deposit onto the riverbed, raising its base elevation. Due to its high sediment load this natural process is particularly rapid for the Yellow River—on the order of 10 centimeters per year. To keep pace, the levees, too, had to be regularly raised.

Eventually, the Yellow River was flowing above its surrounding floodplain. When some levee or dike inevitably failed, the river would rage to lower ground, flooding the valley and occasionally carving out a brand-new path to the sea. Historical records aver that the Yellow River has broken through its levees nearly sixteen hundred times in the past two and a half millennia, flooding thousands of small villages and drowning millions. No other river on Earth has caused more human suffering and death. For this reason, the Yellow River is sometimes called "China's Sorrow."

At least twenty-six times the Yellow River has experienced a major avulsion, etching a new course across the North China Plain. Today, these old riverbeds are visible from space and have been mapped on the ground. They splay hundreds of miles apart, vacillating between north and south as if the river were unable to decide whether to flow north to the Bohai Gulf or southeast into the Yellow Sea.

But not all of these devastating floods occurred naturally. In 1938, the leader of China, Chiang Kai-shek, deliberately unleashed one, with extraordinary consequences to himself, his Nationalist Party, and China's political future.

First, some backstory. Chinese politics in the 1920s were chaotic, with different warlord regimes competing for power. The two political parties with the greatest influence, the Nationalist Party and the Chinese Communist Party (CCP), were violently opposed to each other. For a time, they cooperated under the leadership of Sun Yat-sen, who founded the Nationalist Party and was the first leader of post-dynasty China. But after Sun died of cancer in 1925, the staunchly conservative Chiang Kai-shek navigated his way to leadership of the Nationalist Party and the country. The CCP was led by Mao Tse-tung. Chiang vehemently opposed communism, and in 1927 he organized the execution of thousands of Communists in the Shanghai Massacre, collapsing the uneasy alliance and launching China into a civil war.

Japan capitalized on the disarray of this civil war, invading China's three northeastern provinces, then called Manchuria, in 1931. For several years, Chiang Kai-shek divided his attention between containing the Japanese and fighting Mao's Communists. But by 1937 a series of skirmishes with Japan escalated out of control after an incident at the Marco Polo Bridge (or

Lugouqiao), a strategic span over the Yongding River near Beijing. The Nationalists and Communists agreed to halt their civil war in order to turn and fight Japan, beginning a "War of Resistance" against all-out Japanese invasion in July 1937. It was during this tumultuous time that Japanese bombers sank the USS *Panay* in the Yangtze River, as described in Chapter 3. The Second Sino-Japanese War had begun.

China's resistance did not start off well. Nanking (today called Nanjing), the capital city of the Nationalist Party, was soon taken, and by some estimates more than 300,000 of its people were massacred by Japanese soldiers. Chiang Kai-shek withdrew his government west, to the city of Wuhan, but by May 1938 it too was poised to fall. To lose Wuhan would cost China a critical industrial city and likely the loss of aid from the United States and Britain, who had yet to declare war on Japan but opposed its aspirations in China and colonial Southeast Asia. Given the stakes, Chiang Kai-shek was desperate to halt, or at least slow, the Japanese army advancing upon Wuhan.

He chose to do so by turning the Yellow River into the army's path.

In June 1938, with Japanese forces closing in, Chiang ordered the blasting of Yellow River levees near the city of Zhengzhou, in Henan Province. A breach was successfully begun at Huayuankou, a small village north of the city, just upstream of the river's sharp turn northeast toward the Bohai Sea. On June 9, the levee fully opened and the Yellow River surged out of its elevated channel onto lower ground, heading southeast. A flood of water 100 kilometers wide seethed more than 400 kilometers overland before connecting with the Yangtze and Huai Rivers. The mighty Yellow River that hours before had been heading more than 700 kilometers northeast to the Bohai was now carving out a new course, flowing southeast for 1,000 kilometers toward Shanghai and the Yellow Sea.

In 1938, China's leader, Chiang Kai-shek (Jiang Jieshi), deliberately diverted the course of the Yellow River into the path of a Japanese invasion. The flood temporarily delayed the advance, but it eradicated some 900,000 Chinese civilians and more than 3,000 villages and cities without warning. Chiang Kai-shek's Nationalist Party's initial denial of responsibility, followed by lack of remorse, stood in sharp contrast with Mao Tse-tung's (Mao Zedong's) Communists, who aided the flood victims and resettled farmers into the newly empty Yellow River valley land. Their support ultimately helped Mao win a civil war and establish Communist rule over China.

Several thousand Japanese soldiers drowned in the enormous flood that burst from the blown levee. So, too, did an estimated 900,000 unfortunate Chinese citizens living in its path. Forty-four cities and 3,500 villages were inundated or simply washed away. Four million people were made refugees. Half of the arable land in the provinces of Henan, Anhui, and Jiangsu became unusable, contributing to the great Henan Famine of 1942–1943, in which another three million people died. And because there were no levees or other waterworks along the Yellow River's new course, flooding recurred annually during the summer monsoon rains for the next eight years.

Along the old course, the river simply vanished. Five thousand boats were stranded on dry land. Entire fishing communities had to be abandoned. In total, some 12.5 million Chinese were directly affected by Chiang Kai-shek's deliberately triggered avulsion of the Yellow River.

The tactic slowed the Japanese advance but did not stop the fall of Wuhan. The city was taken four months later by Japanese forces advancing west up the Yangtze River. But the delay did allow the Nationalist government enough time to escape. Western allies, apparently convinced of the sincerity of Chiang Kai-shek's effort to resist Japan, continued to supply aid to China, as did the Soviet Union.

By late 1938, Japan's technologically superior military had stalled against a poorly supplied but massive and determined Chinese population. The Second Sino-Japanese War entered a prolonged stalemate, with Chinese resistance led by the Nationalist government from western China and Communist guerrilla fighters behind Japanese lines in northern China.

Three years of further bloodshed would pass before the United States formally declared war on Japan on December 7, 1941. Japan's armies would go on to win a series of battles across colonial Southeast Asia and advance toward India. In 1942 a critical U.S. victory at Midway Island proved to be a turning point in the war, presaging a hopscotch of gory island battles up the South Pacific, while Britain wrested back its colony Burma.

On August 6, 1945, the U.S. Air Force dropped the first atomic bomb on the Japanese city of Hiroshima, and three days later dropped a second bomb on Nagasaki, killing an estimated 120,000 civilians and hastening the end of the war. Nine days later Japan announced that it would surrender, and did so formally on September 2, 1945, to General Douglas MacArthur, Supreme Commander of the Allied Powers, on the deck of the USS *Missouri* in Tokyo Bay. Seven days later Japan officially surrendered to China in Nanjing. Its surrender of Taiwan took place soon after in Zhongshan Hall, the very building where I would give a keynote address nearly seventy years later.

~

Japan's surrender ceremonies did not end the violence in China. With the foreign armies removed, the truce between Chiang Kai-shek's Nationalists and Mao Tse-tung's Communists soon dissolved, reigniting China's still-unsettled civil war. Chiang's deliberate blasting of the Yellow River's levees would figure importantly in this conflict, and in its final outcome.

For nine long years, the aftermath of the avulsion festered. Initially, China's Nationalist government denied responsibility, blaming the disaster on Japanese bombs. Japan vigorously denied this claim, and within weeks the press had accurately identified the real perpetrator as the Chinese government. Tolerance of civilian deaths was very different in 1930s China than today—it was more common at that time for leaders and the general public to accept collateral damage for the good of the nation—but the government's denials heaped further insult on a populace already infuriated by the sheer scale of the disaster and nearly a million pointless deaths at the hand of its own leadership. Public resentment against the Nationalists grew, and the mood would soon worsen.

Recall that Chiang Kai-shek's deliberately triggered avulsion created a wide, abandoned, dry riverbed where the Yellow River used to flow north. Into this space entered Mao Tse-tung's Communists, who helped to organize and settle as many as half a million peasant farmers, including many of the flood's victims, in this now-empty land crossing parts of Henan, Hebei, and Shandong provinces.

As these new settlements took root, a fight broke out between the Nationalists and Communists over whether or not to restore the Yellow River to its old course. The Nationalist government announced a plan to close the levee breach and turn the river back into its original, pre-1938 channel. Survivors still living in the ravaged areas of the Henan, Anhui, and Jiangsu provinces strongly favored the plan, as the river's new path had devoured much of their farmland and what was left continued to experience repeated and dangerous annual flooding. The CCP opposed the plan, out of concern for the half-million farmers it had helped settle onto the dry riverbed. The Yellow River diversion project became politicized, with Nationalists supporting the river's restoration and Communists opposing it.

After World War II, the Nationalists won international funding for the project from the United Nations Relief and Rehabilitation Administration (UNRRA), a relief agency established to assist rebuilding in countries occupied during the war. Its China Program was one of UNRRA's biggest aid initiatives, second only to its reconstruction work in Germany. However, UNRRA withdrew funding for the project when it became evident that the Nationalists were aggressively pushing ahead without first building necessary new levees and repairing old ones to protect the dry riverbed's new inhabitants.

Indeed, Chiang had political reasons for seeking a fast return of the Yellow River to its old course. Doing so would win support of the flood victims in the Henan, Anhui, and Jiangsu provinces, an area generally sympathetic to his Nationalist government. It would insert a formidable water barrier between two major Communist military bases, the Jin-Ji-Lu-Yu Field Army to the north of the riverbed and the East China Field Army to the south, in southern Shandong. For their part, Mao's Communists demanded more time and resources to build dikes and levees to protect the new settlers, as well as relief funds for those who would inevitably be made homeless by the return of the river. The Yellow River diversion project became even more politicized, amplifying the simmering conflict between Nationalists and Communists and helping to deliver the country into a full-scale civil war.

On December 27, 1946, the Nationalists suddenly released a small amount of water into the old pre-1938 riverbed without warning. It was a signal to the Communists that Chiang Kai-shek was ending negotiations. Immediately afterward, Chiang launched the Strong Point Offensive, a series of military attacks against CCP forces in northern China. Lacking time and supplies and busy defending themselves from the attacks, the Communists abandoned their levee preparations, and the newly settled communities were left unprepared for the river's return.

On March 15, 1947, Chiang ordered that the breach be closed for good. Once more the Yellow River turned, this time back into its old valley, heading northeast. Again, the communities that had settled on the dry riverbed received no warning. Nearly five hundred villages were inundated and hundreds of residents drowned. More than 100,000 people became homeless. While this was a relatively small number of casualties compared to the original 1938 flood disaster, the callousness of the unannounced diversion further damaged the standing of the Nationalist Party in the eyes of the public and became an effective recruiting tool for the CCP.

The Communists turned the new crisis to their advantage. Once again, they organized the affected communities and helped to repair levees and dikes along the reoccupied river valley. This further discredited the governing Nationalist Party and evolved into an anti-Nationalist movement. The sympathy and support that the Communists reaped from the affected rural farmers after the initial catastrophe of 1938 and then the second one in 1947 became an important contributing factor for the CCP's recruiting, fighting, and winning the civil war.

The Nationalists' idea of using the Yellow River to physically divide the Communists' bases proved ineffective. Indeed, the diversion eliminated the water barrier between some Nationalist and CCP territories, giving the Communists further opportunity to expand. An area within the 1938 flood zone became a base where the CCP New Fourth Army adapted guerrilla warfare for fighting in floodplain terrain under the leadership of a commander named Peng Xuefeng. These skills would prove critical in the 1948–1949 Huaihai Campaign fought in the Yellow River floodplain, one of the largest and most pivotal battles of China's civil war.

A series of Communist victories snowballed—the Liaoshen Campaign in Manchuria, the Pingjin Campaign, and finally an advance into South China. With the help of local converts, Communist forces launched the Yangtze River Crossing Campaign, capturing still more territory. In early 1949, Communist forces captured Nanking, the Nationalists' capital. On October 1, 1949, Mao Tse-tung proclaimed the People's Republic of China as a sovereign nation, with himself as its head of state and chairman of the Chinese Communist Party, which rules China to this day.

Chiang Kai-shek and his remaining supporters fled to Taiwan, where the "Republic of China" still governs as a political party. An authoritarian with strong backing from the United States, he would go on to firmly bind Taiwan to the West. He dreamed of the collapse of Chinese Communism and of reunification with the mainland until his death, at the age of eighty-seven, in 1975.

Though the Yellow River floods and their aftermath are not widely acknowledged as turning points in Chinese history, the recruitment of rural peasants to the CCP they incited was an important contributing factor to Communist power at a critical time in the country's evolving civil war. The CCP's mass mobilization and flood recovery efforts, together with popular disgust at the ruling Nationalists' callousness toward the disaster victims, helped the Communist Party attract the recruits it needed to fight and win the Huaihai Campaign and other decisive battles. Chiang Kai-shek's deliberately triggered Yellow River floods helped to align millions in support of the Chinese Communist Party and its ultimately successful quest for the rule of mainland China.

It Came from the South Fork Club

Very early one Christmas Eve, my parents' house caught on fire while we slept. We escaped into the snowy night with no belongings and wearing nothing but our nightclothes. The origin of the fire was eventually traced to a defective heater. My parents had good fire insurance, so they didn't sue the manufacturer. But they could have, because America is a litigious society.

On billboards all across the country, attorneys advertise their services with exhortations such as "Injured? Call now!" In 1992, the U.S. District Court for the District of New Mexico ordered McDonald's to pay $2.7 million in punitive damages to a customer who was badly burned by hot coffee that spilled in her lap. In 2009, Francisco Garcia of the Sacramento Kings was injured after the balancing ball on which he had been exercising burst. The Kings and Garcia sued the manufacturer and distributors for $4 million in salary paid to Garcia while he was recovering, plus $29.6 million in punitive damages. In 2015, Toyota paid $1.1 billion to settle a class-action lawsuit accusing the company of overlooking an important safety feature. Though all of the injuries described in these various lawsuits were unintentional, the plaintiffs still had the unequivocal right to hold the defendants responsible for their damages and suffering.

Lawsuits like these are standard in America because U.S. courts follow the legal doctrine of "strict liability." Strict liability allows a victim to seek reimbursement and punitive damages if someone harms them, even if that harm was accidental. Ignorance or good intentions are not protections against legal action, nor is proof of gross negligence necessary for compensatory damages to be awarded. Strict liability is why so many lawsuits are filed in the United States, and it has greatly improved the safety of products and workplaces in the country.

American jurisprudence was not always this way. In the nineteenth century, hard evidence of deliberate *intent* to cause harm or at least willful negligence was required for a defendant to incur any damages whatsoever. How did the United States, a country founded by fiercely independent religious outcasts, with an easygoing culture of redemption and some of the most forgiving bankruptcy laws in the world, become so tough? The answer traces back to a series of influential flood disasters in England and the United States, culminating with an epic 1889 catastrophe in Johnstown, Pennsylvania.

Strategically located at the confluence of the Little Conemaugh and Stonycreek Rivers (where they form the Conemaugh River) in the resource-rich Allegheny Mountains, Johnstown boomed as a transport hub in the 1830s after the Conemaugh was incorporated into the Pennsylvania Main Line, a new canal system linking Philadelphia and Pittsburgh. To supplement low summertime water levels in the canal, the Commonwealth of Pennsylvania built the South Fork Dam across the Little Conemaugh, fourteen miles upstream of Johnstown, to create an artificial reservoir from which water could be released into the canal system as needed.

The original South Fork Dam was made of earth and stone and was some 70 feet high, 918 feet across, and 220 feet wide at its base, narrowing to 10 feet at the top. It had spillways and relief pipes to allow excess water to escape, and an underground stone culvert with drainpipes so that the reservoir could be drained in an emergency. The dam took years to complete, and by the time it was finished in 1852, the era of canals was giving way to the era of railways. The entire purpose for the reservoir became obsolete, and the state sold off the dam, 450-acre reservoir, and surrounding land.

After its sale, the system fell into disrepair. In 1862, a storm damaged the culvert and breached part of the dam. Neither was repaired. Over the years, the relief pipes were sold for scrap metal. In 1879, after twenty-seven years of neglect by various owners, the reservoir and South Fork Dam were bought by a railroad builder and real estate investor named Benjamin Ruff, who renamed the reservoir Lake Conemaugh and converted the property into an exclusive retreat for millionaires, called the South Fork Fishing and Hunting Club.

Ruff brought in one of his railroad construction gangs to fill in the breach in the dam, using inappropriate materials and building methods designed for railroad beds. The culvert was sealed up, and the missing drainpipes were not replaced. The top of the dam was cut down to widen its surface, reducing the flow capacity of the remaining spillway. Metal screens were hung across this spillway to prevent stocked bass from escaping into the Little Conemaugh River. Some disturbing leaks in the dam were patched up with horse manure and straw. No engineer was brought in to consult on any of these modifications or repairs.

The South Fork Club built vacation cottages and an elegant lodge with forty-seven guest rooms and a grand formal dining hall. For nearly a decade, the idyllic site was a retreat and playground for Pittsburgh's wealthiest families. Its members included Andrew Carnegie, Henry Clay Frick, and Andrew Mellon.

In 1880, an engineer warned the club that the repairs were insubstantial, the dam was leaking, and the lack of any drainage pipe posed an existential threat to the downstream valley and city of Johnstown, home to more than thirty thousand people (including many immigrants from Germany, Wales, and Ireland) and one of the country's leading producers of iron and steel. The South Fork Club ignored the engineer's warning, even after water levels in the Lake Conemaugh reservoir climbed to unsafe levels and the dam began to sag. Pittsburgh's millionaires fished, dined, and picnicked, seemingly unconcerned about the Frankenstein's monster of a dam separating 20 million tons of water from a booming, nationally important steel-producing valley downstream.

On May 28, 1889, a great storm began and the rivers around Johnstown went into flood. Runoff from the hills surrounding Lake Conemaugh began to raise the reservoir's water level even higher. Leaves, sticks, and other debris washed in, plugging the metal fish screens that the South Fork Club had installed across the dam's spillway. Realizing the danger, some locals desperately tried, in vain, to remove or clean the screens. The water level behind the dam was rising a foot per hour when they abandoned their efforts and fled, around 1:30 p.m. on May 31. The water soon overtopped the dam, and at approximately 3:10 p.m. the contents of Lake Conemaugh ripped through the failing structure and surged down the valley toward Johnstown.

The flood wave was moving more than 100 miles per hour, with heights of 50 feet or more, when it struck Johnstown with a force comparable to that of Niagara Falls. Within minutes, the city was obliterated. Rumors about the dam's poor condition had circulated for years, so many residents had already fled for the hills. Those who didn't paid with their lives. The official death toll was 2,209, roughly 10 percent of Johnstown's population. Sixteen hundred buildings were simply swept away. "Everything—houses, engines, everything—just tumbling around in the water," said one survivor, describing the scene.

The bodies were carried off as far away as Cincinnati, Ohio. Skeletons would continue to turn up for the next twenty-two years. The flood threw debris against a railroad bridge, piling up a forty-foot heap of lumber and victims. As the flood waters subsided, still-burning coal embers set the pile afire. The desperate screams of

Gross negligence by the South Fork Club, a fishing retreat for wealthy elites, set the stage for a catastrophic dam failure and resultant downstream annihilation of the booming city of Johnstown, Pennsylvania. Public outrage over the club's exoneration triggered nationwide adoption of strict liability laws, a sea change in American jurisprudence. Today, the 1889 Johnstown Flood still ranks as one of the deadliest disasters in U.S. history, and lawsuits based on strict liability are ubiquitous in the country.

people trapped inside would haunt their frantic would-be rescuers for the rest of their lives. Up to eighty people who managed to survive drowning burned to death in the fiery wreckage.

Within days, Johnstown was swarming with journalists from around the country. Like the 1927 Mississippi River flood and Hurricane Katrina in 2005, the flood's aftermath gripped the country, with newspapers publishing daily updates on the death toll, damages, and relief effort. The Republican U.S. president Benjamin Harrison implored the nation for support.

Donations and volunteers poured into Johnstown to help with the relief and recovery effort. One of those volunteers was a woman named Clara Barton, who had recently started a new disaster-relief organization. She organized volunteers, cleanup teams, and construction crews. Donors sent her mattresses, stoves, shoes, and $3 million in cash. Her success, together with the tremendous level of national publicity, dramatically boosted her fledgling organization, the American Red Cross, in its first major test.

~

As the true cause of the Johnstown Flood became known, the nation was infuriated by the negligence and arrogance of the South Fork Club. Because its members were millionaires, people assumed the club would compensate the victims and rebuild the town. Instead, it denied any culpability and donated only trivially to the recovery effort. This provoked outrage across the country and at one point a furious mob even attacked the South Fork Club's elegant lodge.

A formal investigation concluded that gross negligence by the South Fork Club was directly responsible for the dam's collapse, all 2,209 lives lost, and approximately $17 million (nearly $500 million in 2020 dollars) in property damage alone. This announcement produced still more newspaper editorials and public outcry, with calls for the courts to compel compensation for the victims. But in 1889, U.S. courts had not yet adopted strict liability. Many lawsuits were brought against the South Fork Club, but all failed. Neither the club nor any of its members were ever held liable for the death and destruction that their negligence had unleashed upon the hapless people of Johnstown.

As occasionally happens in America, there was a national awakening. The country absorbed the reality that a small group of millionaires, through their negligence, had destroyed an entire town without consequence to themselves. One of the most influential law journals in the country called for U.S. courts to adopt strict liability, citing the gentlemen's storage of "a vast reservoir of water behind a rotten dam, for the mere pleasure of using it for a fishing pond." For precedent, they looked across the Atlantic to England, which had also been grappling with the very same problem of negligence-induced floods.

The English case was called *Rylands v. Fletcher,* and it would become far more influential than anyone dreamed. It was shaped by at least three flood disasters that took place in nineteenth-century Britain. The first was a catastrophic failure of the Bilberry Dam, a problematic structure built across the River Holme to power textile mills near the village of Holmfirth in Yorkshire. Among many design flaws, its embankment was built atop an actively flowing spring, which eroded it from beneath. An emergency control valve to release excess water from its reservoir was inoperable. The dam collapsed on February 5, 1852, killing at least seventy-eight people and displacing seven thousand others.

A second catastrophe was the collapse of the Dale Dyke Dam across the River Loxley, near Sheffield, on the night of March 11, 1864. The resulting flood raged down the valley in the early-morning hours, sweeping hundreds of people from their beds. This one killed at least 238 people and displaced more than twenty thousand.

Both disasters received wide publicity in England and influenced how the House of Lords decided to rule on the comparatively mundane case of *Rylands v. Fletcher,* in which the plaintiff, Thomas Fletcher, sought compensation for damages from the defendant, John Rylands. Rylands had dammed up a small private reservoir on his property, which then accidentally broke through into an abandoned coal mine shaft and flooded out Fletcher's nearby mining operation. The House of Lords ruled in favor of Mr. Fletcher, and the case of *Rylands v. Fletcher* was set to become a watershed moment in British and American tort law.

The *Rylands v. Fletcher* ruling offered a first precedent for strict liability. In 1886, for example, the California Supreme

Court cited *Rylands v. Fletcher* to establish strict liability in the state after a series of destructive floods and lawsuits involving damage from hydraulic mining of river placer deposits. But it was the Johnstown Flood that solidified this movement. In the wake of the public outcry that followed the disaster, many U.S. states, including the highest courts of Massachusetts, Minnesota, Maryland, Ohio, Vermont, South Carolina, Oregon, Missouri, Iowa, Colorado, West Virginia, and Texas, began broadly applying *Rylands v. Fletcher* to a wide range of industrial hazards. Even the high courts of New Jersey, New York, and Pennsylvania—which had previously rejected arguments based on *Rylands v. Fletcher*—reversed their opposition within a few short years of the Johnstown Flood. In fits and starts, U.S. state courts would go on to apply it in legal decisions all across America.

This legal sea change came too late to help the Johnstown survivors. But, as so often happens after floods, the city was quickly rebuilt, and its population surged more than 40 percent over the next several years.

The death toll from the Johnstown Flood approached that of the September 11, 2001, terrorist attacks in New York, Washington, D.C., and Pennsylvania, placing it among the very worst disasters ever to occur on U.S. soil. Today, the Johnstown Flood National Memorial, run by the U.S. National Park Service on the former site of Lake Conemaugh, is a popular tourist destination. Eerily, United Airlines Flight 93, the fourth hijacked 9/11 plane, in which passengers and flight crew and terrorists fought to their deaths, crashed near Shanksville, Pennsylvania, less than 20 miles from Johnstown, resulting in the creation of a second national memorial nearby.

Like the 9/11 attacks, the unanticipated disaster had lasting influence. The Johnstown Flood of 1889 helped to launch the American Red Cross and, within a few short years, set into motion a tidal shift in societal expectations of responsibility and legal liability in America.

CHAPTER 5

SEIZING THE CURRENT

In 2018, a wave of cinematic excitement washed over the world when Marvel Studios released *Black Panther,* a superhero action film set in the fictional East African country of Wakanda. The movie absolutely crushed box-office records, earning back its estimated $200 million in production costs on opening weekend. For months the public offices of Wauconda, a small town in Illinois, suffered prank phone calls of *Wakandaaaa!* (the movie's signature battle cry) followed by teenage giggles and hang-ups. Within three months, *Black Panther* was the third-highest-grossing movie of all time and a global sensation, earning $1.3 billion by the end of the summer. When the government of Saudi Arabia lifted a thirty-five-year ban on cinemas, the first movie screened in Riyadh's gleaming new theater complex was *Black Panther.* The following year, the film was nominated for seven Academy Awards and won three of them.

The movie was refreshing for at least two reasons. First, its thirty-two-year-old director Ryan Coogler and nearly all his cast members were black, including some of African nationalities. This shattered a long-held Hollywood maxim that films with black casts have a narrow market. Second, *Black Panther* dared to imagine a world in which the most sophisticated, rich, and technologically advanced civilization is a never-colonized African country.

In the film (and in the Marvel comic books on which the film is based), the source of Wakanda's technological power is an otherworldly metal called vibranium having near-magical energetic properties. More realistically, Wakanda also relies on a river. It flows through the center of Wakanda's towering capital city and also creates Warrior Falls, a sacred waterfall where ritual battles for the throne of Wakanda are fought to the death in its cascading pools. It is in these pools that the movie's hero, T'Challa (played by Chadwick Boseman), and antagonist, Killmonger (played by Michael B. Jordan), battle for the right to rule Wakanda and whether the country will abandon its traditional passivity to assert a more aggressive, leading role in the world's geopolitical order.

In the months following the film's release, *Black Panther* drew some comparisons to an actual country in East Africa. It, too, had never been truly colonized, although Italy tried. Comparisons were made between the technology-savvy T'Challa and Emperor Menelik II, who in the nineteenth century embraced modern military technology to repel an invasion of his country. Like Wakanda, it is a proud nation with diverse tribes and a healthy respect for education. It too depends on rivers, and one very important river in particular. And by asserting a more aggressive, leading role over that river, Ethiopia is abandoning its historically passive role in the region's geopolitical order.

The GERD

I first heard about Ethiopia's controversial dam project a few years ago in, of all places, the Arctic. A Norwegian colleague of mine in Bodø, learning of my interest in rivers, asked in a conspiratorial voice if I'd heard the news yet about Ethiopia. This impoverished country, where incomes are among the world's lowest and where horrific famines occurred in the 1980s, was defying the World Bank to build a massive dam across the source of the Nile. No, I did not know about that, I said, but I'd sure like to. Soon after, the project was brought up to me again, this time in a Shanghai pub, by the prominent British environmental author Fred Pearce, who had just written an article about it. It seemed that everyone I knew wanted to talk about one of the most fascinating riparian power struggles of modern times, now unfolding in East Africa.

At the heart of this controversy sits Egypt. Recall from Chapter 1 that to the ancient Egyptians the annual Nile River flood that appeared out of the sunbaked, bone-dry desert was a gift from the gods. Egypt's civilization depended upon the Nile for its very survival, and still does.

Egypt has long been the biggest consumer of the Nile's water, despite being its most downstream riparian state. The country's desire to legally protect this status quo dates back to historical treaties it signed with Sudan, its immediate upstream neighbor, before and after both countries became British colonies. These treaties divided up virtually all of the Nile's water between them. The most recent, the 1959 Nile Waters Agreement, allocates 55.5 cubic kilometers to Egypt and 18.5 cubic kilometers to Sudan annually. But today, there are nine other sovereign nations farther upstream in the Nile River Basin, and their own water needs are unrecognized by these old treaties. A new international agreement that includes all of these countries is badly needed. Yet any reduction in the total volume of water flowing downstream is potentially devastating for Egypt.

The most glaring omission from the old treaties is Ethiopia. In fairness, our hydrological understanding of the enormity of runoff from the Ethiopian Highlands to the Nile River Basin was still quite poor in 1959. But today we know that nearly 90 percent of the water flowing in the Nile River originates there, in one of Africa's largest natural water towers. Thousands of streams

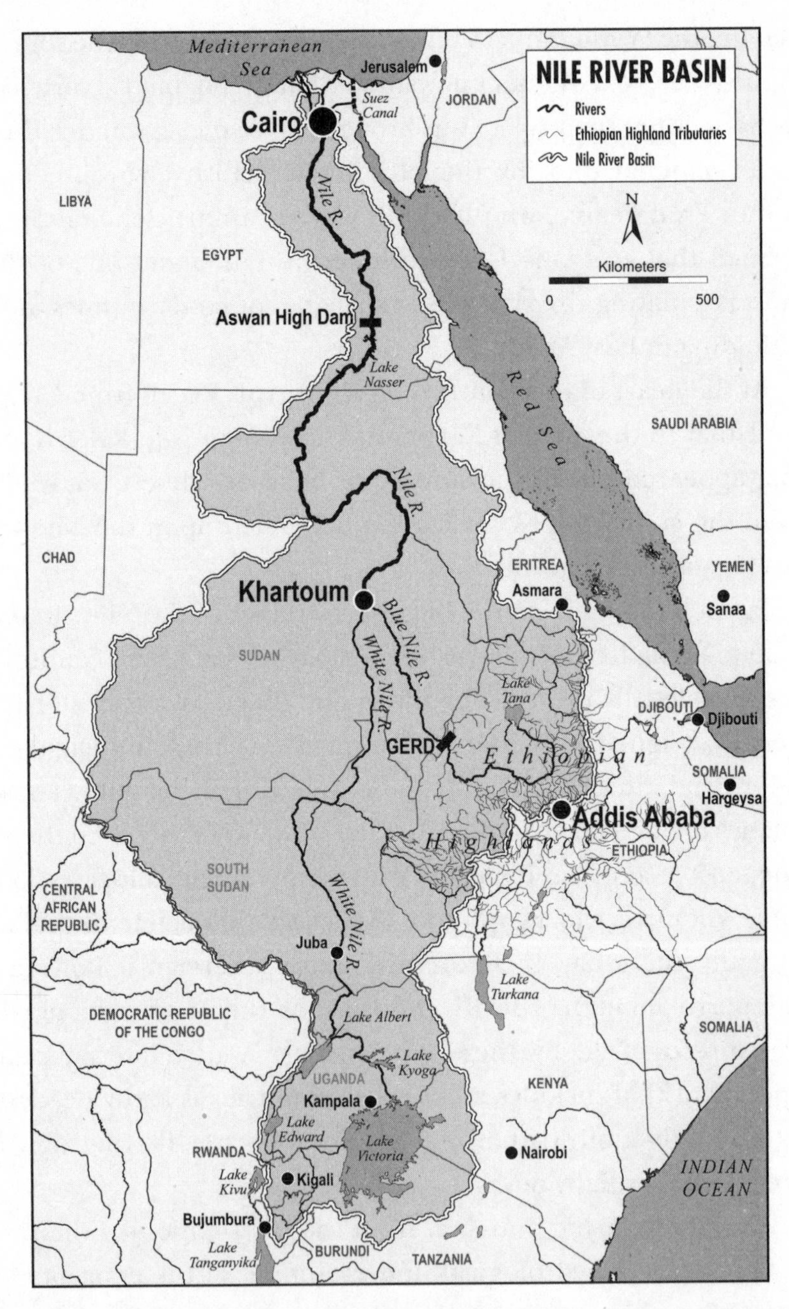

Egypt's existential, millennia-long reliance on the Nile is now threatened by contemporary geopolitics. The headwaters for this vital river comprise eleven sovereign nations, with most of its water sourced from the Ethiopian Highlands. For this reason, Ethiopia's construction of the Grand Ethiopian Renaissance Dam (GERD) is of enormous concern to Cairo.

and tributaries that are headwaters for the Nile emanate from this water tower. The biggest of all is the Blue Nile River.

The Blue Nile is the easternmost of two great tributaries that converge to form the Nile River at Khartoum, Sudan. The other is the White Nile, which has headwaters in six countries (the Democratic Republic of the Congo, Rwanda, South Sudan, Sudan, Tanzania, and Uganda) and includes Lake Victoria. From Khartoum, the Nile flows north to Egypt with only modest water withdrawal by the Sudanese. It courses over one thousand miles through the barren Sahara Desert before being impounded by the Aswan High Dam to create Lake Nasser, a vast artificial reservoir. This reservoir has killed off Egypt's famous annual flood cycle since 1970 but has replaced it with new benefits in the form of a stable water supply and electricity for the well-being of Egypt's rapidly growing population.

The importance of the Nile's water to Egypt cannot be overstated. Hence the country's vehement, almost frantic opposition to Ethiopia's plan to dam the Blue Nile, which supplies at least half of the Nile River's water.

In 2011, Ethiopia announced a detailed plan to build a massive hydropower dam and reservoir across the Blue Nile about 30 kilometers upstream of the Sudanese border. A construction contract was signed with Salini Impregilo, an Italian civil engineering firm, the following year. The groundbreaking ceremony happened on March 24, 2011, just weeks after Egypt's strongman president Hosni Mubarak was deposed in the Arab Spring revolutions that lit up the Middle East. Despite financial and technical setbacks, construction has more or less proceeded ever since. At the time of writing, construction crews are working continuously around the clock and the project is around 70 percent complete (see color plate).

The name of this engineering megaproject is the Grand Ethiopian Renaissance Dam, often shortened to GERD. When completed, it will stand 155 meters tall and 1,780 meters across, making it the largest dam in Africa. It will infill a reservoir of approximately 1,870 square kilometers, rivalling Egypt's Lake Nasser in area. The system will have an installed electrical generating capacity of up to 6,000 megawatts, roughly triple the current production capacity of the Aswan High Dam. Around twenty thousand people will be displaced. The estimated cost of the GERD is nearly $5 billion.

This is an enormous price tag for such a small country: $5 billion is about 6 percent of Ethiopia's gross domestic product and 40 percent of the government's total annual operating budget. Soon after announcing the project, Ethiopia began applying for international financing from numerous sources, including the World Bank, the European Bank for Reconstruction and Development (EBRD), the Export-Import (Exim) Bank of China, and various sovereign wealth funds.

The reaction from Egypt, which views any interference with the Nile River's flow as an existential threat, was punishing and swift. As Ethiopia sought loans overseas, Egypt lobbied ferociously to block any international capital from reaching the project. It sent requests to the United Nations and the African Union asking that the funding requests be denied, citing risks to regional stability. In 2013, a gathering of Egyptian politicians was caught on live television openly discussing sabotaging or bombing the dam. More reasonably, Egypt demanded that the project be halted until an international panel of experts could be convened to study its impacts. Under intense Egyptian pressure, Ethiopia's financing requests to the World Bank and EBRD, and later, China's Exim Bank, were all denied.

But that was not the end of the GERD. After the international loans were refused, Ethiopia turned inward. In 2011, the prime minister, Meles Zenawi, called for Ethiopians to come together and shoulder the cost themselves. His government devised a plan to fund the project in a highly unusual public-private partnership with its own citizens. Ethiopia's government would pay around 80 percent, and Ethiopia's people would pick up the rest. The idea caught on, receiving widespread popular support that snowballed into a national movement.

According to Fekahmed Negash, executive director of the Eastern Nile Technical Regional Office, the GERD "has created a sort of national unity in Ethiopia." Supportive Ethiopians have several options to help fund the project: They can make an outright donation, they can play the national lottery, or they can buy bonds that are either interest-free or yield a modest fixed interest of around 2 percent per year. All government employees and members of the military are expected to donate up to a month of their salary each year, first as grants, then later as payments for bonds. Prominent business owners are also urged to give grants. University students receiving government-funded room and board are encouraged to fast—with credit for the skipped meals earmarked for the GERD. Fund-raising events abound. Indeed, when I met Negash, he said he was running a 5K race the next day to help raise money for the dam. In 2018, an estimated half-million Ethiopians in more than a hundred cities ran a fund-raising race to help support the GERD.

This kind of grassroots fund-raising is happening not only inside the country but throughout Ethiopia's global diaspora. GERD bonds can be purchased in Ethiopian consulates around the world for as little as a few dollars. Curious, I tried to buy a small bond myself and learned that the offer is only available to Ethiopian nationals or foreigners of Ethiopian descent. The bonds cannot be sold or transferred to non-Ethiopians.

Somehow, the GERD has transformed from a piece of utilities infrastructure to a national pride movement. Thousands of Ethiopians travel hundreds of miles to see the dam construction site in a remote, lightly populated corner of their country. The anniversary date of the project's groundbreaking ceremony has become something of an annual celebration.

Where does this fervor come from? It's impossible to imagine anyone in my own country, for example, conceding a month of their salary to fund a power plant or any such public infrastructure. According to Negash, one reason traces back to the early 1980s famines, which killed more than a million Ethiopians in 1984 alone. "It was a national shame," he told me. "Every year, begging for food from the international community...it was a humiliation." Ethiopia also has a long-standing animus with Egypt, over both water and Egypt's backing of Eritrea in past civil and independence wars. Egypt's lobbying of international financiers to reject the GERD was thus unsurprising to most Ethiopians and consistent with a long history of antagonism between the two countries. Viewed in this light, we can see that the GERD has become something bigger for many Ethiopians—a way to flex their country's power over Egypt and the world.

Egypt's lobbying effort backfired. Unfettered by World Bank procedures, Ethiopia is plowing ahead without the battery of external reviews that would otherwise have been required. Some environmental NGOs have lined up against the project, citing opacity and insufficient technical and scientific studies. A nonpartisan group of international scholars convenes intermittently at the Massachusetts Institute of Technology to identify ways to reduce regional tension caused by the GERD and to examine potential engineering risks. The group identified weaknesses in its saddle dam, for example, which could potentially fail and unleash a catastrophic flood on Sudan. Ethiopian officials point to Salini Impregilo's design, which will use concrete to strengthen the suspect structure and its weak underlying bedrock, but Egypt's civil engineers remain skeptical.

For Egypt, the gravest near-term threat of the GERD surrounds its reservoir-filling period. Its reservoir capacity will be enormous; it could diminish or even halt downstream flows in the Blue Nile for years while it fills. An abrupt stoppage of Blue Nile discharge would devastate Egypt.

A long-term threat is that by impounding water during the rainy season and releasing it later, during the dry season, the GERD will tempt Sudan to tap more of its legally entitled water allocation under the 1959 Nile Waters Agreement (much of which, as previously noted, now flows unused into Egypt) to expand its irrigated agriculture. Unlike hydropower generation, which alters the timing of river flow but is otherwise non-consumptive (apart from evaporation), irrigating crops in the desert sharply lowers the volume and quality of water returned to the river. While Sudan is not currently using its water allocation, the GERD's increase in dry-season discharge will make this practicable for Sudanese farmers, denying Egypt its accustomed surplus.

An agreement between Egypt, Ethiopia, and Sudan on how to gradually fill the GERD reservoir, and how to manage its storage collaboratively with Lake Nasser, is badly needed. The 1959 agreement is outdated and unrecognized by the other Nile Basin nations. Some promising cooperative frameworks have emerged—such as the Nile Basin Initiative, formalized in 1999—but nothing resembling a hard water-allocation treaty, signed by all eleven riparian nations, is on the horizon. A framework agreement to equitably share Nile Basin water was signed by Ethiopia, Uganda, Kenya, Tanzania, Burundi, and Rwanda by 2011, but it was opposed by Egypt and Sudan.

With trust, smart water management, and binding commitments not to expand Sudanese irrigation, the downstream impacts of the Grand Ethiopian Renaissance Dam on Egyptian water supply are likely controllable once the reservoir-filling period is over. Clearly, the GERD and the Aswan High Dam will need to work together. The GERD might even increase Egypt's water supply by retaining water that would otherwise evaporate faster from Lake Nasser in the baking Egyptian heat. Some stirrings of cooperation have emerged, including the signing in Khartoum of a declaration of agreed-upon principles by the leaders of Egypt, Sudan, and Ethiopia in 2015. There are also some signs of warming relations between Egypt's president Abdel Fattah al-Sisi and Ethiopia's prime minister Abiy Ahmed.

Meanwhile, the controversies and intrigue continue. In 2018, the GERD's chief engineer, under pressure because of delays and increasing costs of the project, shot himself in Addis Ababa. In 2019, Sudan was thrown into disarray when the country's thirty-year dictator, Omar al-Bashir, was overthrown by a military coup following months of demonstrations in Khartoum. At the time of writing, no agreement between the three countries on how to fill and manage the GERD—let alone a grand transboundary water-sharing treaty among all the Nile River Basin countries—is anywhere in sight.

As in the biblical story of Joseph and Pharaoh, the next seven years could see feast or famine in Egypt, depending on how Ethiopia, Egypt, and Sudan negotiate the arrival of the Grand Ethiopian Renaissance Dam, which is slated for completion in 2022. In the meantime, Ethiopia is projecting power over the region by damming the Blue Nile against the will of international financiers, environmental groups, and its livid, stronger neighbor downstream. *Wakandaaaa!*

The Megadam Century

The Grand Ethiopian Renaissance Dam is just the latest example of a global trend in massive river engineering megaprojects, which has been proceeding for nearly a century.

Throughout human history, countless societies have tapped, diverted, and dammed rivers. The archeological evidence for this is abundant, from long-forgotten waterworks dug across Mesopotamia to the remnant footings of thousands of former millraces in the United Kingdom and New England. However, all previous activities seem almost innocuous compared to the sheer size and power of the big twentieth-century river engineering megaprojects.

The trend began in America during the Great Depression, with a series of government-funded New Deal megaprojects intended to employ people and harness the natural capital of the Colorado, Columbia, Missouri, and Tennessee Rivers for irrigation, energy, and development. Huge structures that were built or started during this time include the Hoover Dam, the Grand Coulee Dam, the Fort Peck Dam, and the Tennessee Valley Authority dam system. These American megaprojects inspired comparable ones in Canada, the Soviet Union, India, and elsewhere, with enthusiastic backing by national governments and the World Bank.

The 1950s and '60s, in particular, saw extraordinary investments in massive reservoir dams and associated hydropower and water supply infrastructure. Major rivers that were dammed during this time include the Angara and Yenisei Rivers in Russia (Bratsk Dam, Krasnoyarsk Dam), the Caroní River in Venezuela (Guri Dam), the Indus River in Pakistan (Tarbela Dam), the Paraná River in Brazil and Paraguay (Itaipu Dam), the Peace and Manicouagan Rivers in Canada (W. A. C. Bennett Dam, Daniel-Johnson Dam), the Sutlej River in India (Bhakra Dam), the Nile River in Egypt (Aswan High Dam), the Volta River in Ghana (Akosombo Dam), and the Zambezi River in Zambia and Zimbabwe (Kariba Dam). Like their American predecessors, these megaprojects had transformative effects on the local economies and settlement patterns in these countries.

Today, a new wave of river engineering megaprojects is rolling through the developing world, many of them even bigger than their twentieth-century forerunners. China's gigantic Three Gorges Dam spanning the Yangtze River stands about 594 feet tall and 7,770 feet long and took some twenty years to complete. Its huge reservoir began impounding the river's discharge in 2006, filling the longest artificial lake in the world, which stretches some 370 miles between Chongqing and Sandouping, roughly the distance from Cleveland to Washington, D.C.

To make way for this long, narrow reservoir, approximately 1.3 million people were evicted from their homes along the Yangtze River and its tributaries. More than 400 square miles of land, thirteen cities, nearly fifteen hundred towns and villages, and numerous archeological and cultural heritage sites were submerged. Together with a monetary cost of more than $50 billion, these sacrifices turned the Yangtze River into a renewable-energy colossus. With an installed capacity of 22,500 megawatts, the Three Gorges Dam is by far the biggest hydroelectric power facility in the world at this time.

In Brazil, construction is under way on the Belo Monte Dam (actually a series of dams and linked reservoirs) to harness the Xingu River, a tributary of the Amazon. The project will displace tens of thousands of indigenous people and badly degrade their fishing grounds, while supplying more than 11,200 megawatts of hydropower capacity to the national grid. In Southeast Asia, numerous dams are planned along the Lower Mekong River, as discussed in Chapter 2. These projects will help to electrify Southeast Asia and bring sorely needed revenue to Laos, at the cost of local livelihoods and ecological damage to one of the greatest freshwater fish baskets in the world.

In the Democratic Republic of the Congo, a truly massive proposed hydropower scheme envisions building a series of low dams across the mighty Congo River. Called the Grand Inga, this megaproject would dwarf even the Three Gorges Dam in terms of cost and installed power-generating capacity. The first phase proposes a new dam and power plant (called Inga-3) on Inga Falls, a great set of rapids down which the world's second-largest river (in terms of discharge) tumbles toward the Atlantic Ocean. These rapids, located about 25 miles upstream of the port of Matadi, have long attracted proposals for hydropower development despite the area's political instability. As of 2019 an international consortium of investors, led by China and Spain (Three Gorges Corporation and Actividades de Construccion y Servicios SA) had submitted a $14 billion bid to proceed with the initial construction of an 11,000-megawatt facility at Inga-3. Current cost estimates for the full Grand Inga approach $90 billion for an installed generation capacity of 40,000 megawatts, nearly double the cost and power of the Three Gorges Dam. If the Grand Inga is ever built, it will generate more than one fourth of the total electricity produced on the African continent today.

Three Inventions That Changed the World

As transformative as the huge twentieth- and twenty-first-century megaprojects are, they are just the latest example of societies engineering rivers to suit their needs. The Three Gorges Dam, Belo Monte Dam, and Grand Inga scheme are modern (if extreme) incarnations of one of three ancient technological marvels: the dam, the diversion, and the bridge.

Bridges are so ageless and ubiquitous that hardly anyone gives them much thought. Surely, early hominids used that simplest of bridges, a log over a stream, just as wildlife and hikers do today. The first known writing of the word *bridge* (the Greek word *γέφυρα*) appears in Homer's *Iliad*, but archeological evidence of their use dates back much further. Two of the oldest known examples were built during the Middle Bronze Age Mycenaean civilization, outside the present-day hamlet of Arkadiko in rural Greece.

These two particular bridges are still standing after more than three thousand years. Their narrow, corbelled arches were assembled by carefully fitting boulders together, and both structures are amazingly well preserved and still usable today. Should you wish to find them, the geolocation of one bridge is 37° 35′ 37.10″N, 22° 56′ 15.21″E; the second, less visited bridge is at 37° 35′ 27.27″N, 22° 55′ 36.30″E, about one kilometer away. These and other Mycenaean bridges served the same purpose in the Bronze Age that road culverts do today: They allowed people in vehicles to blaze down the road without falling into a stream. Mycenaean chariots have since been replaced by cars and trucks, but the concept is the same.

What began as a pragmatic road improvement, of course, became so much more. The Romans perfected the use of fitted stone and concrete to make many arched bridges and viaducts, including masterpieces like the Alcántara Bridge over the Tagus River in Spain. Other early feats of stone include the seventh-century Zhaozhou Bridge over the Jiao River in China, the fifteenth-century Charles Bridge over the Vltava River in Prague, the seventeenth-century Si-o-se-pol Bridge over the Zayandeh River in Iran, and the seventeenth-century Pont Neuf over the Seine River in Paris.

The world's first cast-iron arch bridge was built in 1779 across the River Severn in Shropshire, England. Small, costly bridges made of fitted stone became obsolete as our proficiency with metalwork improved, and societies turned to longer, stronger, cheaper spans made of iron, steel, and reinforced concrete. By the early twentieth century, many iconic bridges in North America and Europe had been built, including the Brooklyn Bridge in New York, the Burrard Bridge in Vancouver, the Golden Gate Bridge in San Francisco, the Chain Bridge in Budapest, the Oberbaum Bridge in Berlin, and Tower Bridge in London.

As more and larger bridges were built, rivers transformed from being barriers to overland travel into concentrators of social and business activity at bridgeheads. River crossings, by their very nature, attract flows of people, transportation, and commerce.

They generate toll revenue and encourage settlement on both banks, thus placing the river at the center of the city's growth. While ferries serve a similar purpose, the permanence, ease, and reliability of bridges encourage construction of permanent buildings on both sides of a river.

The two halves of Paris, a city originally founded on an island in the middle of the Seine, are today integrated by thirty-seven bridges that stitch together two roughly comparably-sized urban areas on either side. Without them, Paris would have grown asymmetrically—only on one side, like the Russian cities of Volgograd and Yakutsk. In fact, most large cities today are bisected by a river running through their cores, an important observation that I shall return to and quantify near the end of this book.

Where rivers define international political borders, bridge crossings often grow paired border towns on either side, like the twin cities of El Paso and Ciudad Juárez, described in Chapter 2. Bridges are strategic in time of war, as we saw in our discussions of Operation Market Garden and the Battle of Sedan in Chapter 3. Some are works of art, like the Dragon Bridge in Vietnam and Helix Bridge in Singapore. Some even have symbolic power.

Take, for example, a recently completed bridge linking the Krasnodar region of southwestern Russia with the Crimean Peninsula. Over the protests of Ukraine and the international community, this highly controversial structure was completed in 2018, after Russia's 2014 invasion and annexation of Crimea, at the time a part of Ukraine. This expensive twelve-mile bridge spans the Kerch Strait, surpassing the Vasco da Gama Bridge in Portugal as the longest bridge in Europe. When it opened, Russia's president, Vladimir Putin, drove the first truck over this figurative and literal bridgehead onto Crimean soil. At a cost of almost $4 billion and international tension, this bridge now provides a direct road link from Moscow to Crimea and is a symbolic as well as physical projection of Russian power.

The year 2018 also marked the fiftieth anniversary of the Nanjing Yangtze River Bridge, one of the first permanent bridges built across the Yangtze. Before its completion in 1968, people and goods had to be floated across the mile-wide river in boats. Even trains were disassembled, put on ferries, then reassembled on the other side, a highly inefficient process. With an upper deck boasting four car lanes and pedestrian sidewalks and a lower deck carrying a railway linking Beijing to Shanghai, the Nanjing Yangtze River Bridge was an important milestone for the country's developing transportation system.

It was also a symbolic milestone, because it was the first major twentieth-century bridge to be designed and built solely by China. Its technologically modern yet distinctly Chinese architecture was a rare source of pride in the country during some of the worst horrors of the Cultural Revolution. The Nanjing Yangtze River Bridge appeared on cups, pencils, shoes, bicycles, and other products sold across China, and adorned propaganda posters promoting Mao Zedong Thought (Maoism).

Today, the Nanjing Yangtze River Bridge remains a culturally important icon in China; it received a $160.7 million restoration in honor of its fiftieth birthday. Elsewhere in the country, many of the world's most technologically exciting bridge spans are being built. Recent marvels include the Duge Bridge, the world's highest, with a deck soaring 1,854 feet (565 meters) over the Beipan River to link the mountainous provinces of Guizhou and Yunnan, and the Yangsigang Bridge, the world's longest double-deck suspension bridge, over the Yangtze River in Wuhan.

Artificial Rivers

Alongside dams and bridges, the third river-engineering marvel that changed how societies use rivers was the *diversion.* Like the bridge, its origins are prehistoric. Archeological evidence of stream diversions predates even the great hydraulic societies—for example, in northern Iraq, where early proto-farmers experimented with redirecting stream courses onto their fields at least nine thousand years ago. Recall from Chapter 1 that the great civilizations along the Nile, Tigris-Euphrates, Indus, and Yellow River valleys built elaborate waterworks of ditches and canals to divert water onto crops. Between 850 and 1450 CE, the Hohokam civilization built a system of irrigation canals near present-day Phoenix, Arizona. Today, the world's largest contiguous irrigation system, on the Indus River floodplain in Pakistan, uses more than 60,000 kilometers of diversion canals.

Put simply, the entire premise of irrigated agriculture rests on water diversion, an idea that has been discovered and rediscovered through the ages. Plop children in a sandbox with a running garden hose and some sticks to dig with, and they will soon mimic their ancestors by inventing the same thing.

Diversions are also useful for floating boats. No one knows where or when the world's first river channel was canalized, but a reasonable guess is Iraq, during the golden age of Sumerian city-states described in Chapter 1. These early Mesopotamian cities traded by watercraft along the shifting channels and cutoffs of the Tigris and Euphrates Rivers. Owing to the rivers' high sediment loads, their preferred routes would have required regular dredging, a first step toward digging the area's numerous short canals—faint traces of which are still visible from space today.

The modern-day Suez Canal connecting the Mediterranean and Red Seas is one of the most important shipping shortcuts in the world. However, it is not the first canal to have been dug through this narrow isthmus. The first was cut long ago at the directive of Egyptian pharaohs, who used tens of thousands of slaves to hand-dig a 100-mile-long "Canal of the Pharaohs," using little more than bronze shovels. This ancient shipping route, which splayed eastward off the Nile along the course of a natural *wadi* called the Wadi Tumilat, was used for centuries. Its approximate course is still visible today in satellite images as a verdant strand of green (today the corridor is lined with irrigated cropland) on the Nile Delta. It veers sharply away from the Nile, traveling east from Zagazig through Ismailia and sharply turning south to reach the Red Sea, along a route that is now part of the Suez Canal

The world's longest shipping canal, extending some 1,100 miles, was built in China. Called the Grand Canal or Beijing-Hangzhou Canal, it was completed in 609 CE after extending and connecting some waterways and even older canal segments dug a millennium earlier, in the fifth century BCE. The Grand Canal created a long, critically important inland shipping corridor linking the Yangtze and Yellow Rivers to each other and to a series of cities and villages from Beijing to Hangzhou. This access was vital for shipping and commerce and became a historically integrating force in China. It is still in use today.

In France, completion of the Canal du Midi in 1681 allowed boats to fully traverse the country for the first time. This remarkable canal linked the Mediterranean Sea to the city of Toulouse, and from there (via the Garonne Canal and the Garonne River) to the Atlantic Ocean. With its 150-mile route of tunnels, aqueducts, and nearly a hundred locks, the Canal du Midi was one of the great wonders of the world when it was constructed and continues to be used, mainly by pleasure craft, to this day.

Shipping canals took a great technological leap forward in 1761, with the invention of the first fully artificial canal in Lancashire, England. The Duke of Bridgewater, who owned some productive but landlocked coal mines about ten miles northwest of Manchester, wished to find an economically viable way for the city's booming textile industry to access his coal. No natural waterway existed near his mines, and slogging coal on pack-horses was prohibitively expensive. Inspired by the Canal du Midi, he hired an engineer named James Brindley to invent some sort of canal-like solution.

Brindley designed an ingenious, fully artificial canal to connect the duke's coal mines directly to Manchester. The canal originated underground, inside the mines, before crossing *over* the River Irwell on an aqueduct bridge and continuing on to the city. Barges were loaded inside the mines and then pulled by packhorse on a towpath alongside the canal. Brindley's design even helped to drain the mines of groundwater, thus turning a chronic problem into a source of water to help fill the canal and keep the barges afloat.

Within a year, the Bridgewater Canal had slashed the price of Manchester coal in half. The Duke of Bridgewater became rich, and his success inspired a canal-building boom in England. Aided by Brindley's conceptual breakthrough, aqueduct bridges, and the coal-fired steam shovel, many new overland canals were dug over the next forty years. These linked, for example, the Severn River to the Mersey and Thames Rivers, and the Mersey to the Trent. These and other new canal networks vastly improved the speed and cost at which raw materials and commerce flowed around central England, helping to fuel its rapidly industrializing economy.

Germany also embraced the technology, and the country continued digging important shipping canals well into the late nineteenth century. The Dortmund-Ems Canal, a 167-mile waterway connecting the industrial heartland of the Ruhr River valley to the North Sea, was completed in 1899. Together with the Wesel-Datteln Canal, the Datteln-Hamm Canal, the Mittelland Canal, and the Elbe-Lübeck Canal, it helped create an arterial transportation and shipping network linking Germany's industrial centers with the Elbe and Rhine Rivers and the Baltic and North Seas. The shorter but crucial Kiel Canal, a 61-mile waterway connecting the North Sea at Brunsbüttelkoog (at the mouth of the Elbe) to the Baltic Sea at Kiel-Holtenau, was completed in 1895. It was subsequently modified many times and remains an important part of Germany's inland transportation system today. The 106-mile Main-Danube Canal, completed in 1992, links the Rhine and Danube Rivers, creating a 2,200-mile artificial waterway permitting ship passage between the North and Black Seas.

Across the Atlantic, England's canal-building boom inspired a similar burst of hydraulic engineering in America. One of the most effective projects was the Erie Canal, which helped to open up the country by linking the Atlantic Ocean, via the Hudson River, with the Great Lakes and points west. When the Erie Canal opened in 1825, it was 40 feet wide, 4 feet deep, and 363 miles long. It had eighty-three locks to raise its water levels the nearly 600 feet in elevation needed to float barges between the Hudson River and Lake Erie. It had eighteen aqueduct bridges to soar them over troublesome ravines and rivers. Each barge could carry up to 30 tons of freight or passengers, towed by horses plodding along a towpath next to the canal. This vital new transportation route would undergo still more deepening and extensions in subsequent years, creating a broad, arterial canal network across the region.

By the mid-nineteenth century, the superiority of rail technology was apparent and the golden age of canals drew to an end. Germany continued to expand its network of artificial inland waterways for decades, but the last very long English canal was completed in 1834. In Pennsylvania, a certain dam and reservoir that had been built on the Little Conemaugh River to release water into the state's canal system was sold and fell into disrepair, setting the stage for its eventual purchase by a club of elite Pittsburgh millionaires and the catastrophic Johnstown Flood described in Chapter 4.

For nearly a century, a canal-building frenzy had gripped Britain, continental Europe, and the United States. While technically an evolutionary improvement on the centuries-old practice of deepening and canalizing natural waterways and using locks to raise and lower boats, the idea of striking out overland with a fully artificial channel and raised aqueduct bridges was a radical technological advance. Like the railroads that supplanted them and the high-speed interstate highways that would follow rail, canals were a disruptive advance in transportation technology that energized the industrialization of Europe and the westward growth of the United States.

Many of these old canals are still used today, including China's Grand Canal, France's Canal du Midi, and America's Erie Canal. Indeed, today's 524-mile New York State Canal System connects rivers and lakes across the state and into Canada. Now a vibrant recreational destination with more than two thousand attractions and tourist facilities, it is traveled by pleasure boats and cyclists seeking a water's-edge echo of a time when artificial rivers were the highways of America.

An Irishman Builds Los Angeles

More than a century has passed since William Mulholland, an Irish-born civil engineer and the first superintendent of the Los Angeles Department of Water and Power (LADWP), hatched a plan to surreptitiously buy up the land and water rights of a remote river and divert it south to his city. His target was the snowmelt-fed Owens River, flowing through a bucolic rancher's paradise nestled between the Sierra Nevada and Death Valley more than 200 miles north of Los Angeles.

Mullholland's clandestine water grab diverted virtually all of the snowmelt runoff cascading down the eastern flank of the Sierras into the Owens River. The move prompted rebellions and bombings by the valley's furious residents and would one day inspire Roman Polanski's classic 1974 movie *Chinatown,* starring Jack Nicholson and Faye Dunaway. It also recast the image of Los Angeles as a bare-knuckled city that would grow by whatever means necessary. By 1913, the heist was complete, and Mulholland presided over the grand opening of the Los Angeles Aqueduct, a 233-mile canal and pipeline that diverted the Owens River into the San Fernando Valley. At the time, it was the longest aqueduct in the world. "There it is. Take it!" Mulholland famously exclaimed. And Angelenos did, triggering a new population boom in what is now the second-largest urban agglomeration in America.

The LADWP's power play on a distant river was the first of several diversions of faraway rivers into Southern California. In 1919, the U.S. Geological Survey first proposed diverting water from Northern California's Sacramento River south through the San Joaquin Valley to reach the drier, southern part of the state. By 1931, designs for a great north-to-south water-transfer plan were on paper. Seven years later, during the depths of the Great Depression, the federal government launched this plan by damming the Sacramento River near Redding. Today the Shasta Dam is just one component of the Central Valley Project, a vast federal system of reservoirs, hydropower dams, and canals serving farmers and population centers throughout California's Central Valley.

By the late 1940s and the 1950s, a postwar population boom was building pressure for another river diversion plan to send water to Southern California, funded by the State of California. Like the federal Central Valley Project, the state system would build a vast interlinked network of dams and reservoirs to store and shuttle water and also generate the enormous amount of electricity needed to pump it throughout the area and over the Tehachapi Mountains into Los Angeles. In 1960, a bitterly fought

California bond measure was narrowly passed to begin funding this sprawling north-to-south river diversion plan, blandly named the State Water Project.

The referendum pitted one half of the state against the other, with Southern Californians viewing the State Water Project as essential for growth and Northern Californians vehemently opposed to any raid on their water resources. The measure barely passed, and to this day northerners harbor a brooding dislike of southerners, which can be traced all the way back to this voter-endorsed water grab. The grudge is one-sided, with a diaspora of Angelenos cheerfully traveling north to enjoy wine-tastings and the lovely city of San Francisco, while their Patagonia-fleeced hosts dourly wonder how anyone could stand to live in L.A.

But Angelenos are not put off by any of this. They are a happy people, blessed with comfortable weather, a vibrant culture, and far more water than their local climate deserves. The 1960 bond measure built the huge Oroville Dam across the Feather River in north-central California. Its reservoir, Lake Oroville, became the pumping heart of an expansive network of reservoirs, hydropower dams, power-generating stations, aqueducts, tunnels, and pumping stations that circulate water south for two thirds of the length of California and some 700 miles of built infrastructure. (In 2017, a near-catastrophic failure of the Oroville Dam's spillway threatened the state's water supply and 200,000 residents downstream, who were forced to evacuate—concerns remain about the long-term safety of the dam, even after a $1.1 billion repair job was completed in 2019.) The State Water Project irrigates three quarters of a million acres of farmland and delivers water to more than 27 million people in Northern California, the Bay Area, the Central Coast, the San Joaquin Valley, and Southern California.

The project's biggest customer by far is the Metropolitan

Water District of Southern California, a mammoth conglomerate of public agencies that oversees the delivery of water to 19 million people living in Los Angeles, Orange, Riverside, San Bernardino, San Diego, and Ventura Counties. Metropolitan owns and operates nine reservoirs, sixteen hydropower facilities, four of the world's largest water treatment plants, and the Colorado River Aqueduct, which carries water 242 miles from the Colorado River to Southern California. It is also a world pioneer in exciting new groundwater storage and "toilet-to-tap" recycling programs, a subject I will return to in Chapter 8. Metropolitan was the driving force behind the State Water Project's proposed "California WaterFix" project, a plan to bore two tunnels under the Sacramento–San Joaquin River Delta east of San Francisco to further improve the conveyance of Sacramento River water to Los Angeles. Prior to its cancellation, for environmental reasons, by California's governor Gavin Newsom in 2019, Metropolitan had committed $10.8 billion to California WaterFix, making it the project's single-largest financial backer.

California's Central Valley Project and State Water Project, together with regional river diversions like the Los Angeles Aqueduct and Hetch Hetchy pipeline/reservoir system connecting Yosemite National Park to San Francisco, are major river diversions that have thoroughly replumbed the state. San Franciscans became river diverters in 1913, when they won the right to dam the Tuolumne River and turn its Hetch Hetchy Valley, one of the loveliest in Yosemite National Park, into a water-supply reservoir. That flexing of urban power triggered such furor that it launched an environmentalist movement, spearheaded by pioneering conservationist and author John Muir. The protestors lost, and the Hetch Hetchy now provides most of San Francisco's water supply—meaning that the diverted Tuolumne River now provides natural capital and human well-being to one of the most vibrant and innovative cities in the world.

Grand Diversions

Today, our demand for water is motivating plans for new river mega-engineering projects at a scale and pace the world has never seen.

Around 4 billion people—almost two thirds of the living human race—experience severe water scarcity for at least one month out of the year. Of these 4 billion, some 900 million live in China and 1 billion in India. Other large populations experiencing water scarcity are living in Bangladesh, Pakistan, Nigeria, Mexico, and certain dry western and southern U.S. states. In recent years several large cities have experienced acute water shortages, including São Paulo in Brazil, Chennai in India, and Cape Town in South Africa.

These problems will worsen. Even setting aside, for the moment, adverse impacts of climate change, human demand for fresh water is expected to rise to more than 6 trillion cubic meters per year by the middle of this century—more than 50 percent higher than today. India, in particular, faces a daunting challenge of tripling its fresh water supply by 2050 in order to meet the needs of its rapidly growing and industrializing population.

To help alleviate these pressures, the coming years will see the most elaborate river diversion schemes ever imagined. Local water diversions have been with us for millennia, but huge interbasin river diversions are a different beast. First, they require immense engineering infrastructure and modification of the landscape. Second, by transporting river water over great distances, they encourage development and population growth in places that simply cannot survive on local water sources alone.

This practice began in twentieth-century California and has now gone global.

Consider just three grand interbasin river diversion schemes that are now either in planning stages or already under way: China's South-to-North Water Diversion Project, Africa's Transaqua Project, and India's National River Linking Project.

China's South-to-North Water Diversion is currently the largest existing interbasin river diversion scheme in the world. Its overall purpose is to shunt water from the country's wet south to its arid north. It will do this by diverting water from the mighty Yangtze, as it flows eastward from the Tibetan Plateau to the East China Sea, into three long, north-flowing canals in western, central, and eastern China (see map next page).

The idea is an old one. Besides floating boats of grain, the ancient Grand Canal was also intended to divert water from the Yangtze to the dry northern part of the country. Plans for the current project date back at least fifty years, to the rule of Mao. Its master plan was approved by China's State Council in 2002, and the project broke ground later that year.

The eastern route, which taps the Yangtze River at Yangzhou city and incorporates the Grand Canal in its flow north to Tianjin, was completed in 2013. The central route diverts water from a reservoir on the Han River (a major tributary of the Yangtze), flows north to the Huai River basin, then tunnels under the Yellow River to eventually end up in Beijing. This 804-mile central route, completed in 2014, forced the relocation of a third of a million people. It is now delivering water to more than 50 million and is responsible for some 70 percent of Beijing's water supply.

The third, western route is still in the early planning stages. It will divert water from three tributaries of the upper Yangtze near the Tibetan Plateau, dumping it into the headwaters of the oversubscribed Yellow River. This final phase of the project may be completed by 2050.

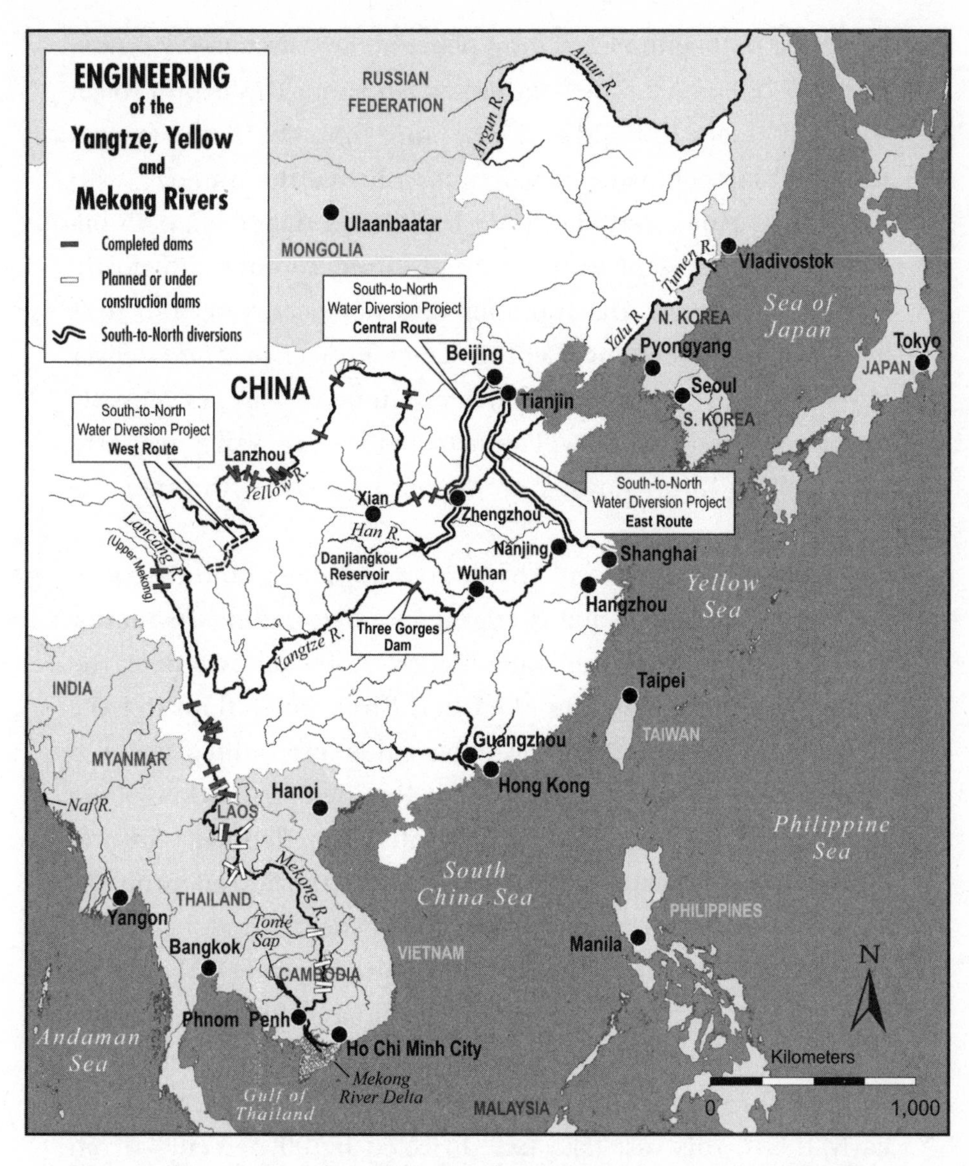

Rivers in China and Southeast Asia are being engineered at a massive scale. Highlighted here are China's South-to-North Water Diversion Project (with east, central, and western routes), shunting water from the Yangtze River basin northward; the Three Gorges Dam; and a cascade of new dams completed or planned down the Mekong River through China, Laos, Thailand, and Cambodia.

With a half-century building plan and an estimated cost of at least $77 billion, China's South-to-North Water Diversion Project will take longer to build and cost more than the Three Gorges Dam. When complete, it will interconnect the Yangtze with the Yellow, Huai, and Hai River basins and divert some 45 billion cubic meters of water annually from southern to northern China. To put that number into perspective, California's State Water Project and Central Valley Project *combined* deliver fewer than 14 billion cubic meters annually. China's massive, multipronged diversion will approximate the flow of a new, artificial Yellow River, flowing from south to north through the country.

In central and western Africa, an ambitious grand diversion scheme called Transaqua is edging closer to reality. Its master plan envisages diverting water nearly 1,500 miles out of the Congo River basin and into the Chari River, which flows to Lake Chad. It would dig a long navigable canal and build a series of hydropower dams in the Democratic Republic of the Congo, the Republic of the Congo, and the Central African Republic. Around 50 billion cubic meters of water would be transferred annually, roughly half of it through an 830-mile artificial canal.

One of the motivations for Transaqua is saving Lake Chad. Once a thriving freshwater ecosystem and a major source of livelihood for millions, Lake Chad has suffered severe desiccation from local irrigation diversions and declining rainfall. Since the early 1960s, this vast lake has shriveled by 90 percent—from 22,000 square kilometers to less than a thousand. The resulting losses of fish, cattle, and crops has created food insecurity and bleak socioeconomic conditions, making the region vulnerable to extremist political movements. Boko Haram, a religious militant group notorious for mass abductions of high school girls, has established a stronghold insurgency in northeastern Nigeria where the shrinkage of Lake Chad has created a truly desperate situation. While the proposed diversion of Congo water would not restore Lake Chad, it could stabilize its area at around 7,500 square kilometers while irrigating up to 70,000 square kilometers of farmland in Cameroon, Chad, Niger, and Nigeria.

Saving a piece of Lake Chad is only one objective of Transaqua. The project's boosters point to its hydropower dams, which would bring badly needed electricity to the region. They applaud its fully navigable waterway, which would help ten mostly landlocked African countries gain better access to one another. They envision a grand new development corridor along this river diversion, with collateral benefits for the region's transportation, agriculture, energy, and industry.

In 2018, Transaqua was the focus of a major summit in Abuja, Nigeria, attended by international backers and enthusiastic presidents from Chad, the Central African Republic, Gabon, Niger, and Nigeria. The project seems to have particularly inspired the Nigerian president, Muhammadu Buhari, who hosted the summit. A series of recommendations and next steps were identified, including the creation of a $50 billion international investment fund to be steered by the African Development Bank. Feasibility studies for Transaqua are already being funded through China's huge Belt and Road infrastructure investment initiative. The social and environmental costs of Transaqua remain poorly understood but are under study. After years of languishing, support for this grand river diversion scheme may be gathering momentum in Africa.

~

The mother of all interbasin river diversion megaprojects is the National River Linking Project (NRLP), now advancing in India. If fully implemented, it would reconfigure dozens of mountainous river headwaters in the Himalayas and connect dozens more lowland rivers with each other. Its overall ambition is no less than to reconfigure the flow of rivers around the entire Indian subcontinent.

Like the other grand diversion schemes described so far, the NRLP aims to divert water from rivers located in wet places—particularly northeastern India, which receives up to fifty times more rainfall than arid parts of the country—into rivers located in dry places. After decades of discussion and study, the project's planners have categorized India's rivers into those typically having a water "surplus" versus those having a "deficit." Long artificial canals and tunnels would transfer water from surplus rivers to deficit rivers.

Hydrologically speaking, this binary distinction grossly oversimplifies the realities of river flow seasonality and how the water is actually used. Basins deluged by monsoon rains produce high discharge but still experience water scarcity during the dry season, for example, and better water management would alleviate shortages even in dry areas. But public support for the NRLP has swelled thanks to sustained backing from India's prime minister Narendra Modi and a recent decision by the Supreme Court of India ruling the project to be in India's national interest.

The NRLP's master plan consists of two broad strategies. The first is to impound and divert runoff between neighboring headwater tributaries of the mighty Ganges and Brahmaputra Rivers. This part of the plan, called the Himalayan Rivers Development Component, would both impound water for later use and divert flow from one watershed to the other. It would be achieved by creating numerous large reservoirs and up to fourteen "links" (aqueducts and tunnels) cutting through topographic headwater divides in the rugged Himalayas of northern India, Nepal, and Bhutan.

The second strategy is to interconnect the many large lowland rivers flowing over the Indian subcontinent. This part of the master plan, called the Peninsular Rivers Development Component,

The Nilometer at Roda Island, Cairo, operated from 861 to 1887, making it one of the youngest examples of this ancient Egyptian technology. By tracking the progression of the annual Nile River flood, Nilometers helped pharaohs to maximize crop yields and tax revenues from their population. *(Lithograph by Louis Haghe, c. 1846)*

Near El Paso, Texas, the Rio Grande is used to define the political border between Texas and New Mexico (top) and between the United States and Mexico (bottom). Along the U.S.-Mexico border, the swiftly flowing American Canal parallels the river (bottom right), creating a perilous barrier to migrants seeking undocumented entry into the United States. *(Laurence C. Smith)*

In 2017 a violent purge by the Myanmar military drove more than 700,000 Rohingya, a Muslim minority, across the Naf River border into Bangladesh. Their fate remains unsettled. *(UNHCR/Roger Arnold)*

A real-world *Apocalypse Now* was experienced by Richard Lorman and thousands of other Vietnam War veterans from the United States, South Vietnam, and North Vietnam who fought along the complex waterways of the Mekong River Delta. The U.S. Mobile Riverine Force operated a brown-water navy of vessels specially modified for riverine warfare, including modified "Zippo" flamethrower vessels used to attack Viet Cong positions concealed along the banks. *(Richard Lorman)*

In defiance of Egypt and the World Bank, Ethiopia is building the Grand Ethiopian Renaissance Dam (GERD) across the Blue Nile River, a major tributary of the Nile. The GERD, photographed here in 2018, has become a source of national pride in Ethiopia and represents a major assertion of regional power. *(Gedion Asfaw)*

In his inaugural 1970 State of the Union speech, Republican U.S. president Richard Nixon announced a costly, sweeping plan to fight water pollution in America and create a new federal agency to define and enforce science-based national pollution standards. Later that year, he created the Environmental Protection Agency (EPA) through executive order. John McCormack, the Democratic Speaker of the House seated behind Nixon, appears aghast. *(Richard Nixon Presidential Library & Museum)*

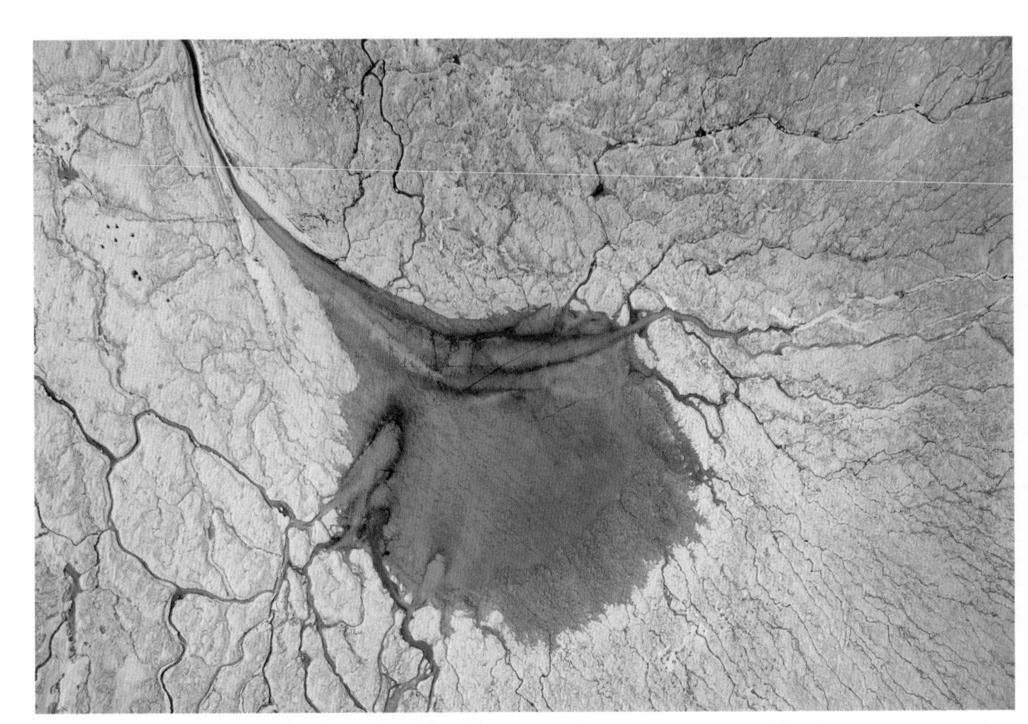

The southwestern side of the Greenland Ice Sheet experiences extensive melting in summer, creating supraglacial lakes, streams, and rivers, as well as moulins that drain meltwater from the ice sheet to the ocean. This process directly contributes to global sea-level rise and is now under intense scientific study. For scale, see the tents of one of the author's field expeditions in the upper left corner of this aerial photograph, acquired by a fixed-wing scientific drone. *(John C. Ryan)*

The late Dr. Alberto Behar was an accomplished robotics engineer at the NASA Jet Propulsion Laboratory, whose passion was inventing autonomous, low-cost sensors to collect scientific data in extreme environments. Here he prepares to launch an unmanned vessel into a supraglacial meltwater river flowing on the Greenland Ice Sheet. *(Laurence C. Smith)*

Rice field fisheries offer unique and sustainable aquaculture for the rural poor by exploiting seasonal flooding cycles to grow fish alongside crops. During the rainy season, these Cambodian rice fields become inundated and fill with spawning fish. As the floodwaters subside, the fish retreat into artificially dug catch ponds, where they are harvested with nets (top) or via draining (bottom). *(Laurence C. Smith)*

Jeffrey Kightlinger (at left), the general manager of Southern California's Metropolitan Water District, presides over the inauguration ceremony for an exciting new water recycling program that will soon serve residents of Los Angeles and Orange Counties in Southern California. Its water source, partly visible in the background, is one of the largest wastewater treatment plants in the region. *(Laurence C. Smith)*

For nearly a century, the Glines Canyon Dam impounded the Elwha River, blocking the passage of Pacific salmon and other migratory fish species. Within days of the final blast demolishing it, Chinook salmon were spotted swimming past this location. Removal of aging dams is a growing trend in North America and Europe, helping to restore rivers to a more natural, free-flowing state. *(John Gussman)*

Artistic rendition of the forthcoming Surface Water and Ocean Topography (SWOT) satellite mission, scheduled for launch in 2022. A joint collaboration between NASA (USA), CNES (France), CSA (Canada), and the UK Space Agency (United Kingdom), the SWOT satellite will track water level changes in the Earth's millions of rivers, lakes, and reservoirs, radically improving our ability to monitor freshwater resources globally. *(NASA)*

Hudson Yards, the costliest urban redevelopment project in U.S. history, is rising over a 28-acre railyard on the Hudson River in lower Manhattan. The $25 billion mix of residential apartments, stores, and public greenspace is scheduled for completion by 2022. *(Laurence C. Smith)*

A New York City wedding ceremony at the end of Pier 61, a former shipping dock now repurposed as an event space, jutting into the Hudson River in lower Manhattan. *(Laurence C. Smith)*

Shanghai is creating abundant greenspace and public parks along its Huangpu riverfront. The Bund, the city's historic esplanade, was revitalized in 2010 and anchors a long-term plan to redevelop both shores of the river. *(Laurence C. Smith)*

involves digging sixteen long canals radiating in all directions around the country. It is expected to commence soon with a diversion called the Ken-Betwa Link Project, which, if it survives challenges by environmental opponents, will consist of a slew of reservoirs, dams, and a 137-mile-long canal linking the states of Uttar Pradesh and Madhya Pradesh.

The sheer scale of India's proposed National River Linking Project is unparalleled, far exceeding even China's South-to-North Water Diversion Project in ambition. Its earth-moving excavations alone would make it the single-largest construction project ever carried out on the surface of the Earth. Fully implemented, the $168 billion project would dig more than 15,000 kilometers of canals and tunnels and transfer 174 billion cubic meters of water per year. It would generate 34,000 megawatts of installed hydropower capacity and increase the area of irrigated land in India by more than a third. Prime Minister Modi calls the NRLP a national dream and has consistently supported it since the start of his administration.

If completed, the NRLP's rerouting of river flows across India would have profound effects on riparian ecosystems and the lives of hundreds of millions of people. It would alter economic development patterns and change livelihoods. It would disrupt fisheries and facilitate the spreading of invasive species, waterborne pollution, and disease. Because rivers are deeply mystical in India, it would disrupt some long-standing religious and cultural practices. Even avid supporters of the National River Linking Project admit that it will damage the environment and displace people, but they argue that these harms will be outweighed by the project's benefits to human well-being and economic growth. These include improved food security and flood protection, as well as new development opportunities for energy, navigation, and urban water supply in what will soon be the most populous country on Earth.

Grand Bargains

The societal benefits of river-engineering megaprojects—reliable water supplies, electricity, flood protection, and economic growth—come with hard costs. These include displaced communities, damage to fisheries and riparian ecosystems, lost navigation access, and heavy taxpayer subsidies. Growing scientific and public understanding of these costs has stiffened resistance to new river megaprojects in the developed world. In the United States and Europe, the boom days of huge dams are over, with more interest in decommissioning and removing obsolete structures than in building new ones (we will return to this in Chapter 7). But in the developing world, the benefits are often deemed worth the costs. A recurring refrain is that the broader societal and economic benefits of river megaprojects outweigh their local environmental and human damage. That argument is made in Brazil today for the Belo Monte Dam, just as it was made in California a century ago for the O'Shaughnessy Dam that turned Hetch Hetchy Valley into a water-supply reservoir for San Francisco. If the city's subsequent success is any example, the argument may well be true. But there is no denying that the tolls are immense.

This is especially true of the huge reservoir dams, which can hold years of a river's flow and submerge entire valleys. Whole towns disappear beneath the waves, and the river's ecology is severed. Confused fish pile up below the dam, blocked from their spawning grounds. Behind the dam, the water stagnates, warms, and loses its oxygen. Pollutants and sediments wash into the reservoir and settle on the bottom, burying the drowned stump forests and cultures of a previous time. The river's discharge becomes controlled, with the timing and volume of its flow regulated by dam operators and electricity-demand pricing cycles. Stripped clean of sediment, the limpid water that is eventually released attacks its own channel, scouring the riverbanks and floodplains for miles downstream.

China's Three Gorges Dam generates nearly twenty times more electricity than California's Hoover Dam and protects 15 million people from deadly Yangtze River floods. These are enormous societal benefits for a deserving region. But to fill its reservoir, more than a million people were evicted and their communities were inundated. The dam has increased water pollution and waterborne diseases, and has triggered new geological hazards in the form of small earthquakes and landslides. Jida Wang, a professor at Kansas State University, discovered that the dearth of sediment in water released by the Three Gorges Dam has eroded the Yangtze riverbed for hundreds of kilometers downstream. This channel down-cutting has deprived the surrounding wetlands and lakes of water, causing them to desiccate. Entire species of fish have disappeared. Of the finless porpoise, a rare freshwater cetacean endemic to the Yangtze, only a thousand or so remain.

The socioeconomic benefits of river megaprojects also tend to turn out differently than supporters predict. An irony of big developing-world hydropower projects is that the new electricity they generate is seldom sent to the rural poor who most need it. A common refrain uttered by every Ethiopian official I have spoken to about the Grand Ethiopian Renaissance Dam is that three out of four Ethiopians lack electricity. Left hanging in the air is the implication that these are the very people who will gain electricity after the dam's construction. Perhaps they will, but not if they keep living where they are now. Ethiopia's lack of national grid infrastructure and the GERD's remote location make selling its electricity to neighboring countries far more sensible. This is not to say that the revenue won't help poor Ethiopians, but the link between this huge river megaproject and improved living conditions for the rural poor is an indirect one.

If anything, big reservoirs have an urbanizing effect—by displacing farmers from the inundated river valleys into cities and extractive-industry sites, the true destination of most of the generated money and power.

We can learn from the damages wreaked by the great river-engineering megaprojects of the twentieth and early twenty-first century. There are hopeful signs that the next generation of big dams being built across the world's last free-flowing rivers in China, Southeast Asia, Latin America, and Africa may use somewhat different technologies than projects of the past. These include so-called run-of-water dam designs, which generate electricity without impounding huge reservoirs, as well as strategies that allow some sediment to pass downstream. Modern dams also incorporate better fish ladders, screens, and other innovations to mitigate their devastating impacts on fish spawning and migration.

As described in Chapter 2, international river agreements have become mainstream. In 2014, the United Nations Watercourses Convention entered force. Cooperative river-governance agreements are beginning to tackle not just water allocations but pollutant and ecological problems as well.

None of these innovations can eliminate the environmental and human tolls of river-engineering megaprojects. But as we shall see next, sometimes we do learn from our mistakes.

CHAPTER 6

PORK SOUP

On March 22, 2017, an ultra runner named Mina Guli ran a marathon along the Colorado River near Las Vegas, Nevada. The following day, she did it again. And then again, again, and again.

After five marathons in five days, she flew to Brazil and ran six marathons in six days along the Amazon River. Then she flew straight to Melbourne, Shanghai, Cairo, and London to run marathons along the Murray-Darling, Yangtze, Nile, and Thames Rivers. By May 1, she had run forty marathons in forty days along six major rivers in six countries.

Mina Guli is a water advocate. The goal of her 1,049-mile "6 River Run" was to raise awareness of the degraded environmental states of these and other rivers due to pollution and water withdrawals. Her passion inspired others to run alongside her and generated considerable media publicity. "Clean, safe, accessible water for all is the most pressing issue our world faces today," she blogged after her marathon tour. For her next awareness campaign, called #RunningDry, she ran 100 marathons in 100 days, all in the name of water advocacy.

Mina Guli is not the first individual to convert personal athleticism into water activism. In the United Kingdom, the late Roger Deakin, an environmentalist author and filmmaker, swam his way across the country, bringing attention to the plight of

Ultra runner and water activist Mina Guli ran forty marathons in forty days alongside six important rivers in 2017—the Colorado, Amazon, Murray-Darling, Yangtze, Nile, and Thames—to draw attention to environmental problems of excessive diversion and pollution of rivers. *(Kelvin Trautman)*

Britain's polluted rivers. In the United States, the environmentalist Christopher Swain has been swimming in polluted rivers since 1996, including the Columbia, Hudson, Mohawk, and Charles Rivers. He swims through sewage, pesticides, and industrial pollutants in some truly horrible places, like Newtown Creek in Brooklyn, New York, which drains a toxic Superfund site. As he swims, he collects data on water quality and his own personal physiology and then shares it with researchers. Like Guli and Deakin, he hopes his actions are bringing public attention to the pressing environmental problems that plague many of the world's rivers today.

Human waste, industrial toxins, and large-scale water withdrawals blight rivers all over the planet. The Ganges, the world's most revered sacred river, is so polluted with fecal coliform bacteria and chemical effluents that it poses a health risk to the millions of Hindu pilgrims who bathe in it as a spiritual rite of passage. Construction of hydropower dams across the Ganges' upper tributaries have halved its discharge in some areas. In 2019, some 150 million people descended upon the river near Prayagraj (formerly Allahabad) to bathe in its sacred waters for a Kumbh Mela festival. Yet despite ongoing pledges by Prime Minister Modi to clean up the Ganges, untreated sewage and factory chemical waste continue to be pumped into the river.

As a scientist, I too have encountered some nasty water. The most unenjoyable fieldwork I have ever done was collecting water samples for my master's thesis. At the coaxing of my professors at Indiana University, I would stand in a dry creek bed during severe thunderstorms to sample the water quality of the flash floods washing off the surrounding hillsides. Whenever the National Weather Service issued a severe thunderstorm watch, I dropped everything and rushed out to my field site in a big Ford Bronco owned by the State of Indiana, loaded up with sampling bottles and sediment nets. I slept on top of the Bronco's roof so the first fat drops of rain would strike my face and wake me.

The wildest storms always came in the dead of night. Donning chest waders and rubber gloves, I would jump into the creek bed with my nets and bottles and peer upstream, half blinded by the Bronco's glaring headlights and glittering sheets of rain, waiting for a flash flood to materialize. I vividly recall one prolonged streak of lightning that lit up the twisted, hellish landscape, freezing into my memory the view of a flood wave rushing at me. A moment later, it crashed through my legs, the water rose, and my long night of dunking sample bottles and sediment nets began.

The work was nasty because these were no ordinary hills and no ordinary creek. The surrounding landscape looked like Tolkien's Mordor, heaped with brooding piles of black, crumbly material called "gob"—shale laced with coal dust that was crushed into chips the size of a thumbnail. Not a blade of green grew on the stuff, which looked burnt but wasn't; the black sooty appearance was intrinsic to the material itself. It was the cast-off spoil from an old abandoned mine that had stripped coal out of the ground a quarter-century earlier.

Mining produces a lot of rock waste. When this waste ("spoil") is piled on the ground, air and rainwater quickly oxidize its freshly exposed pyrite and other minerals, producing a toxic, brightly colored leachate called acid mine drainage. It oozes into the nearby creeks, turning them yellow or orange and sharply lowering the water's pH, killing everything. My particular research project was at an old coal-mining complex called Friar Tuck, one of many such abandoned mine complexes pitting the lovely hills of southwestern Indiana and polluting its groundwater, streams, and the Wabash River.

The gob piles were filthy and acidic, and they would sometimes spontaneously combust. Their soil water was so acidic that a new life-form, a heat- and acid-loving microbe called *Thermoplasma acidophila,* was discovered living near a smoldering underground fire at the site. During intense rains, the loose shale chips would slide en masse down the hillsides, smothering the stream. Even the storm runoff turned instantly acidic, chapping and staining my skin. The research was exciting, but to this day, I dislike the rain.

That old Indiana mining complex is just one of many thousands of abandoned mines ruining water quality around the world. Acid mine drainage seeps out of shafts, spoil piles, and tailings ponds. The leachate spreads toxic amounts of elements such as lead, chromium, manganese, aluminum, and arsenic into the surrounding groundwater and streams. An estimated 20,000 kilometers of waterways around the world are known to be seriously impaired by acid mine drainage, and the true global number is surely much higher.

A Super Fund

In the United States alone, thousands of abandoned mines, industrial sites, military facilities, and old toxic waste dumps still poison streams, rivers, and groundwater. The worst of the worst are designated as federal Superfund cleanup sites—scourges that take decades to decontaminate. One of the first Superfund sites was Love Canal, a short waterway cut into the east bank of the Niagara River in the border town of Niagara Falls, New York.

Situated about five miles upstream from Niagara Falls, the canal was originally conceived as a hydropower scheme. That project was abandoned in 1910, and the partially dug trench was turned into a toxic waste dump. The Hooker Electrochemical Company threw more than 21,000 tons of chemical waste into the hole, then in 1953 capped it off with clay and sold the land to the local school district for a dollar. A tract of homes and a school were then built on the site.

By the late 1970s, corroded drums were poking out of backyards and noxious chemicals were seeping into basements. Local residents were suffering unusual numbers of miscarriages, birth defects, leukemia, and other serious ailments. In 1978, the medical problems at Love Canal were reported on the front page of the *New York Times*. President Jimmy Carter declared a state of emergency, and more than two hundred families were evacuated. A second emergency declaration in 1980 led to the evacuation of 750 more families.

In 1983, Love Canal was designated one of the initial 406 Superfund cleanup sites. More than two decades would be required to decontaminate it. Today when you fly into Toronto, you can still see the canal's rectangular gash from 30,000 feet in the air, but it's no longer so badly contaminated, thanks to the Superfund program.

The Superfund program was created in 1980, when the U.S. Congress enacted a deeply significant law called the Comprehensive Environmental Response, Compensation, and Liability Act, or CERCLA. It granted federal authority to stop or prevent dangerous releases of pollutants into the environment and levied a tax on chemical and petroleum industries for a trust fund (the Superfund) to pay for cleanup of the worst contaminated sites. The legislation was passed by amending an earlier bill called the Resource Conservation and Recovery Act (RCRA), signed into law in 1976 by the Republican president Gerald Ford, which built a framework for managing hazardous and nonhazardous solid wastes in America (and is the main federal law governing the disposal of solid wastes in the country today). The RCRA, CERCLA, and the Superfund revenue stream were critical firsts in a series of progressive environmental laws passed by Republican and Democratic administrations alike throughout the 1970s and '80s.

One catalyst for this stance was the petroleum-slicked surface of Ohio's Cuyahoga River, which spontaneously burst into flames in 1969. A photograph of the lifeless, blazing river was featured in *Time* magazine, sparking public outcry and inspiring generational songs like "Burn On" (Randy Newman) and "Cuyahoga" (R.E.M.). It was not the first time the river had burned, but the public mood had changed. Together with the Love Canal mess and the publication of Rachel Carson's *Silent Spring* (1962), this burning river helped launch a period of strong political opposition to pollution in America.

Back then, the Republican Party had a very different stance on environmental matters than it does today. In January 1970, in his first State of the Union speech, President Richard Nixon laid out his vision of a new decade of environmentalism.

"Occasionally there comes a time when profound and far-reaching events command a break with tradition," Nixon began. "This is such a time." He went on to outline an ambitious plan to regulate and reduce pollution in America, including a $10 billion nationwide water-cleanup program at a time when the country's total gross domestic product was only $1 trillion per year (today, the equivalent GDP percentage would come to nearly $700 billion). "Clean air, clean water, open spaces—these should once again be the birthright of every American.... The program I shall propose to Congress will be the most comprehensive and costly program in this field in America's history," Nixon said.

Such a speech would be unfathomable for a member of Nixon's party today. While studying archived Presidential Library photographs of Nixon delivering it to Congress, I noticed that John W. McCormack, the Democratic Speaker of the House who was seated on the dais behind him, appeared aghast (see color plate). But Nixon would go on to demand billions of dollars for water treatment facilities to clean up the nation's polluted rivers. He proposed national standards for air quality and lower tailpipe emissions from cars. He proposed a tax on leaded gasoline, safeguards against oil spills, and banned the dumping of effluent into the Great Lakes.

A top priority of his administration was to create a powerful new federal agency to help bring about a safer, cleaner country. It would establish and enforce national standards for pollution. It would conduct basic research on how pollutants affect public health and the environment, and develop new ways to reduce them. It would use science and hard data to develop sound policy recommendations for the president and Congress to act upon. The name of this new agency would be the Environmental Protection Agency, and Nixon signed the EPA into existence, by executive order, on December 2, 1970.

Within a decade, the EPA was implementing and enforcing the Clean Water Act, the Clean Air Act, the Toxic Substances Control Act, and the aforementioned RCRA, CERCLA, and Superfund. Thanks to these laws, the past half-century has seen sweeping improvements in what was once a terribly polluted country. The Cuyahoga River is home to beavers, bald eagles, blue herons, and more than sixty species of fish. A symbolic milestone was passed in 2004 when Love Canal was delisted after twenty-one years of cleanup. It is one of 413 Superfund sites that have been certifiably restored. At the time of writing, there are still 1,337 left to go.

This overall trend of tightening pollution control in America reversed sharply in 2017, when the newly inaugurated president Donald J. Trump appointed Scott Pruitt, an avowed critic of the EPA, to head the agency. As attorney general of Oklahoma, Pruitt had sued the EPA fourteen times. Over the course of his sixteen-month tenure he worked to dismantle many EPA norms and regulations, including the rollback of a Waters of the United States rule protecting wetlands and waterways. He aggressively downsized the agency's staff by offering career staffers financial incentives to retire early, then leaving their positions unfilled. Within a year more than 700 employees had left the EPA and Pruitt was proposing to further reduce his agency's workforce from nearly 15,000 employees to fewer than 8,000.

Before Pruitt could accomplish all his goals, he was forced to resign, submerged under a tide of ethics complaints. His successor, Andrew Wheeler, carried on many of his policies more effectively by avoiding the scandals of his predecessor. These have included refusing to resume Superfund dredging in the Hudson River, despite its sediments and fish still containing dangerous polychlorinated biphenyl (PCB) pollutants, and eliminating federal protections for nearly 20 percent of all rivers and streams and half of all wetlands in the country.

This recent weakening of the U.S. Environmental Protection Agency threatens to undermine a half-century of American progress in water-pollution problems. Fifty years ago, the pumping of sewage and industrial sludge into a river was a common sight in the country. The creation of the EPA, and the bipartisan and public support it subsequently received, launched a water-quality improvement trend that survived administrations of both political parties and supported the well-being of millions of Americans.

December 2, 2020, will mark the fifty-year anniversary of President Nixon's EPA and the fortieth year of the Superfund program. Will the next half-century see America's past half-century of accomplishments expanded? I hope so, but in the year 2019 few anticipate a sweeping national plan to reduce greenhouse-gas pollution, for example, as visionary as the Clean Water Act, the Clean Air Act, or CERCLA. To find a country with environmental aspirations like those of 1970s America, we must turn now to China.

China's River Chiefs

After three decades of breakneck industrialization, China has plenty of stories like the Cuyahoga River fire. "Pork Soup," the title of this chapter, derives from a cynical Chinese joke about the tap water in Shanghai after thousands of decaying pig carcasses were discovered floating down the Huangpu River, the city's main water supply, in 2013. A few years before that, a 50-mile-long slick of toxic chemicals, including benzene and nitrobenzene, showed up in the Songhua River, which supplies water for the capital city of Harbin, in Heilongjiang Province.

China's nascent Love Canals await in places like Guiyu, in Guangdong Province, where years of recycling "e-waste"—old computers, printers, mobile phones, and other discarded electronics, mainly from Western countries, have seriously contaminated the water and soil. Common practices at Chinese salvage sites included burning printed circuit boards or cooking them over a grill to liberate electronic components, and dissolving components in open-pit acid baths to extract metals. Low-value plastics and residues were incinerated in open pits or simply dumped in fields and rivers.

These e-waste recycling centers and waste dumps are now badly contaminated with PCBs, PAHs (polycyclic aromatic hydrocarbons), flame retardants, and toxic heavy metals such as lead, cadmium, copper, and chromium. At Guiyu, where a dust of these metals settled over the town, rice grown in nearby paddies contains unsafe levels of lead and cadmium. These sites join thousands of dumps and waterways contaminated in more conventional ways by mines, factories, and petrochemical plants.

Like America's in the 1970s, China's leaders and citizens are growing intolerant of polluted water and air. The country has largely stopped accepting e-waste from foreign countries, and operations at Guiyu, along with some three thousand other unregulated recycling sites, are mostly shut down. On October 18, 2017, Xi Jinping, the president of China and general secretary of the Chinese Communist Party, delivered a major speech to the nineteenth National Congress of the CCP that bore a striking similarity to Nixon's 1970 State of the Union address: "We must realize that lucid waters and lush mountains are invaluable assets... and cherish the environment as we cherish our own lives," he said before a rapt audience. "We will adopt a holistic approach to conserving our mountains, rivers, forests, farmlands, lakes, and grasslands, and implement the strictest possible systems for environmental protection." He proposed stricter pollution standards, new regulatory agencies, and creation of a new "environmental governance system in which government takes the lead." The new entities will be charged with monitoring water and air pollution levels and enforcing new, tougher environmental standards and laws in a centralized way.

Sound familiar? These are pretty much the same targets Nixon laid out forty-seven years earlier, when he proposed sweeping new environmental protections, national pollution standards, and the creation of the Environmental Protection Agency. In the coming years, Xi Jinping's China—whose legislature scrapped presidential term limits in 2018—will see a centralized consolidation of environmental pollution-control policy under a Chinese version of the EPA. The implementation will be different, but the country is bending a new path toward a cleaner future.

Xi Jinping's speech was but one salvo in China's increasingly serious war on water pollution. Using a uniquely Chinese blend of decree and political appointments, the central government has been steadily advancing the idea of a Chinese "ecological civilization," a concept in which material wealth grows alongside a harmonious coexistence with nature.

Improving the state of China's polluted rivers, streams, lakes, and canals is a key step toward this goal. Take, for example, the government's recent creation of a new kind of public employee, called a "river chief," to patrol and enforce environmental laws along natural waterways. A huge number of river chiefs have been hired across the country, with an estimated 200,000 appointments in 2017 alone. The positions are vertically integrated across different levels of government, and if harm befalls a particular river chief's waterway, that person is held personally accountable. If this 200,000-plus inspector army delivers on its stated intent, it will dramatically improve the country's river-pollution problems. Most of China's water-quality problems come not from lack of laws and regulations but from local officials and manufacturers ignoring them.

President Xi's plans for the mighty Yangtze are another example of China's multifaceted effort to clean up its waterways. As we saw in Chapter 5, this river is heavily diverted, impounded by the Three Gorges Dam, and polluted along much of its length. But in 2016 and 2017, President Xi declared that the Yangtze River would become a special economic zone prioritizing ecology and "green development" over traditional manufacturing. Heavy industrial and chemical facilities within 1 kilometer of the Yangtze or its tributaries were outlawed. Within these "ecological redlines," no new projects will be allowed and many existing facilities will be removed.

These new rules represent a radical departure from past development policy in China, and even won high praise from International Rivers, a discerning international NGO based in California, which is dedicated to protecting rivers and riverine communities around the world. If China succeeds, some of its new practices and technologies may spread to other countries within its sphere of international development ambitions. China's Belt and Road Initiative (BRI), in particular, will fund many new infrastructure projects in Central Asia, including dams and water treatment facilities. China's potential transformation from one of the world's heaviest polluters to an emissary of waterway protection would be marvelous for the world, and it is less far-fetched than it sounds.

Streams of Malady

If you enjoy good wine, Renaissance chateaux, and French country cuisine, I recommend a trip to the rolling vineyards and enchanting historic towns of the Loire Valley in central France. The region's delicious white and rosé wines taste even better when sampled inside one of the valley's many cool, musty caves in which the barrels are stored. Meandering through this loveliness is the Loire River and its many tributaries, which created much of the valley floor and its fertile vineyard soils.

These rivers are dotted here and there with little hamlets and villages. Some years ago, my wife and I passed a pleasant week in one of them, named Trôo. Our early evening routine, before settling into a three-hour dinner, was to stroll along the right bank of the Loir River (a tributary of the Loire) where it bends lazily into the village. A stone pier juts out into the water there, and some fish became accustomed to our nightly gift of baguette bits. It all felt so timeless, with the emerald river gliding past waving fields, buzzing bees, and picturesque chateaux.

Pleasant it was, but the timelessness was an illusion. The fish slurping down our breadcrumbs were common carp, an invasive species capitalizing on a serious degradation of water quality throughout the Loire Valley. The Loir River's emerald-green color came from the explosive growth of chlorophyll-rich algae, fueled by nitrate fertilizers seeping into it from one of the most intensely cultivated landscapes in France.

Nitrogen and phosphorous are plant food, and they trigger algal growth in water just as readily as crop growth on land. The resulting blooms turn rivers and lakes bright green and deplete their dissolved oxygen content, a process called eutrophication. When the algae die, they settle to the bottom, where they are devoured by microbes, depleting even more oxygen in the process. Oxygen-loving fish, such as brown trout and barbel, disappear and are replaced by less desirable species tolerant of dank water, like black bullhead, tench, and common carp. With peak concentrations of chlorophyll-*a* exceeding 150 milligrams per liter, the valley's majestic Loire River is one of the most eutrophic rivers in the world.

Choked with algae and depleted of oxygen, such fertilizer-enriched rivers also damage the coastal oceans they flow into. Oceanic "dead zones" are spreading from river mouths and estuaries all over the world. The bottom waters of these dead zones suffer from hypoxia, a condition of acute oxygen depletion (usually defined as less than 2 milligrams of oxygen per liter). Sea life essentially suffocates, triggering confusion and death of benthic life—organisms living on or near the ocean floor. These include commercially important species, such as blue crab and flounder, as well as the mollusks and shellfish that pelagic fishes eat.

More than four hundred dead zones covering more than 245,000 square kilometers have been identified, and their number and extent are growing. Most began forming in the 1960s and '70s, following the advent of nitrogen fertilizers in the late 1940s. One of the largest dead zones forms in the Gulf of Mexico each summer, offshore of Louisiana and eastern Texas. It is fueled by the Mississippi River's discharge, which now has nitrate concentrations up to eight times its preindustrial level. Much of this nitrate comes from Iowa, where it washes off farm fields into the Missouri and upper Mississippi Rivers. In 2017, this particular dead zone grew to a record-breaking area of nearly 9,000 square miles, roughly the size of New Jersey.

Other growing dead zones include Chesapeake Bay and Long Island Sound in the United States, the Elbe Estuary and Kiel Bay in Germany, the Loire and Seine River estuaries in France, the Thames and Forth estuaries in Britain, the Yangtze and Pearl River estuaries in China, and the Gulf of Finland. Because industrial fertilizers are a critical component of the world's current food production system, coastal dead zones spreading from river mouths are a serious marine pollution problem that is expected to worsen in the future.

~

Another commonly overlooked type of river pollution comes from pharmaceuticals. Drug compounds pass through our bodies and are excreted in our urine. People also have a bad (if well-intentioned) habit of flushing old prescription drugs down the toilet. Today's wastewater treatment facilities neither test for pharmaceutical compounds nor have the physical ability to remove them. This means that endocrine-disrupting hormones, antibiotics, antidepressants, and other restricted substances pass right through our bodies and sewage treatment plants into rivers, and from there into the broader ecology of the world.

This is causing some alarming endocrine disruptions in riverine life. In 2005, most of the male smallmouth bass in the Potomac River downstream of a Washington, D.C., wastewater treatment plant were discovered to be carrying female egg cells (oocytes), a mutation attributed to estrogen pharmaceutical compounds flushed down the toilets of Washington. Likewise, a widespread occurrence of intersex fish in United Kingdom rivers has been traced to steroidal estrogen compounds passing through sewage treatment plants. A whopping two thirds of all rivers in the Thames River catchment are believed to be at risk of aquatic endocrine disruptions from this problem.

Antibiotics such as sulfamethoxazole, used to treat urinary tract infections and bronchitis, are also showing up in rivers. Antibacterial chemicals, such as triclosan, are added unnecessarily to hand soaps, body wash, and the like. In general, antibiotics are used too freely by people and fed too liberally to farm animals. As they pass through our sewage treatment plants and into natural waterways, they encourage the evolution of new strains of antibiotic-resistant bacteria with unpredictable effects on ecosystems and public health.

A third commonly overlooked source of river pollution might be sitting in your shower or medicine cabinet right now. Many exfoliating cream or gel skin products use plastic microbeads as an abrasive agent to scrub away dead skin cells. Made of synthetic polymers, these tiny spherical balls are virtually indestructible and too fine to be filtered out by water treatment plants. After dutifully scrubbing away your dead skin cells, they flow merrily through the water treatment process out into our rivers and seas. Small fish, mistaking the particles for food, gobble them up, thus moving the compounds into the food chain. While increasingly banned, skin products containing plastic microbeads are still sold in much of the world.

Microbeads are a particularly pernicious form of a broader epidemic of microplastic pollution of the world's rivers, oceans, and yes—even our treated drinking water. The long-term public health consequences of digesting microscopic particles of plastic is currently unknown, but they are suspected of having carcinogenic and endocrine-disrupting properties. In general, bottled water contains an order of magnitude more of the stuff than tap water.

There are some simple personal choices one can make to help reduce the flow of pharmaceuticals and plastics into rivers and the broader environment. Use skin products containing natural exfoliators—like oatmeal, walnut husks, and pumice—rather than plastic microbeads. Take antibiotics only when essential, and finish the entire prescription when you do. Most pharmacies accept old pills for disposal. At a minimum, keep unfinished medicines sealed in their bottles and throw them into trash headed to a landfill, rather than flushing loose pills down the toilet. Drink tap water. All of these precautions help to lower the growing flood of synthetic compounds entering our rivers, oceans, and the bodies of living things.

The Riviera of Greenland

Since I complained earlier about the nastiest fieldwork I've ever done, it seems only fair to also describe the coolest.

Imagine the world's largest waterslide park. There are rushing chutes everywhere, roiling with aqua-blue water. The chutes come together into bigger ones until there is just one humongous waterslide. It froths past you, some sixty feet wide, barreling mindlessly toward a thundering hole somewhere off to your left. A cloud of white spray rises like smoke from the distant hole. You hear and feel its rumbling beneath your feet.

Now imagine that the waterslides are not made of elevated fiberglass but carved into the surface of the ground. Imagine that the ground is made not of soil but of crunchy white ice. The thundering hole leads not to a swimming pool but to the bottom of the Greenland Ice Sheet. There are certainly no happy shrieks of children. The only sounds you hear are the panting of your own breath, the dragging of a rope clipped to your body, and the distant rumble of the moulin.

This fantastical scene describes the melt zone of the Greenland Ice Sheet, which I have been studying with field expeditions and satellite remote sensing since 2012. Unlike the Antarctic Ice Sheet, large areas of Greenland's ice surface melt during the summer months. This melting is especially widespread in southwestern Greenland, near a small town called Kangerlussuaq (sometimes called Søndre Strømfjord, its former Danish name).

Kangerlussuaq started off as an American air base built during World War II. In 1992, the airport and base were decommissioned and turned over to Greenland, an increasingly autonomous territory of Denmark that is gradually disentangling itself from its former colonial master. Today, Kangerlussuaq Airport (SFJ) is Greenland's only international airport and an aviation hub for the country. Should you ever visit Greenland (which I strongly encourage) from Copenhagen, you will fly right over the melt zone I have described. Be sure to reserve a window seat and bring a good camera, because you've never seen anything like it in your life.

Nicknamed "the Riviera of Greenland," this area of Greenland is sunnier and drier than most. In June and July, sunlight pours down like warm honey nearly twenty-four hours a day, melting away the winter snow and exposing the darker bare glacial ice that lies beneath. This bare ice surface melts down several centimeters per day, and becomes pitted, dirty, and soaked with meltwater. Millions of little rivulets join together like twigs, then branches, and finally a roaring trunk—a supraglacial river—that first meanders and then rages over the ice, flowing at some of the fastest measured river velocities on Earth.

Using satellite mapping, my students and I have discovered that virtually all of the hundreds of supraglacial rivers in this area flow into moulins (meltwater tunnels in the ice). From there, the water escapes under the ice sheet toward its edge—and the ocean. The loss of ice from Greenland is currently raising global sea levels by about 1 millimeter per year—roughly a third of the total rate of global sea-level rise—and is expected to increase even further in the future. The supraglacial rivers we study are ground zero for much of that water.

The meltwater also pools into hundreds of luminous blue supraglacial lakes, dotting the vast white surface like jewels (see color plate of our field camp in Greenland). While beautiful to behold, they are dangerous: Each year, many of these lakes suddenly open up a moulin at their bottom. Within hours, the lake empties, its impounded meltwater plummeting down into the ice sheet and headed toward the sea.

The striking visual impact of Greenland's supraglacial rivers and lakes was one reason why the *New York Times* embedded their Pulitzer Prize–winning photographer Josh Haner and staff reporter Coral Davenport with one of my field expeditions in 2015. Our scientific objective was to test the accuracy of climate models that are used to predict future melting and sea-level rise caused by Greenland's melting. We achieved this by measuring discharge in a large supraglacial river, while simultaneously using drones and satellites to map out its upstream watershed. By comparing our discharge measurements with the simulations of climate models, we can learn how trustworthy their predictions are.

As my team strung cableways over the river and launched data-transmitting floats into the blue torrent, Haner conducted one of the *New York Times*'s first acquisitions of original aerial video by drone. His footage was spectacular. The newspaper would later incorporate it into a multimedia piece titled *Greenland Is Melting Away,* which won a Webby Award in 2016. The following summer, *The New Yorker* magazine sent Elizabeth Kolbert, a Pulitzer Prize–winning writer, to our field camp. Kolbert impressed us all by enthusiastically helping to unload the helicopter and pitch tents on the melting ice. She spent a night with us, then flew home to write the thoughtful article "Greenland Is Melting," about scientists and foreign interlopers in Greenland and the island's continuing power struggle with Denmark. When our scientific findings were published in 2017, the *New York Times* used some of our own drone footage to create a new multimedia piece titled *As Greenland Melts, Where's the Water Going?* Collectively, the three high-profile articles provided the general public with stunning images and insights about this important melting ice sheet and its bizarre system of supraglacial rivers that few people even knew existed.

As these blue rivers rush over the ice sheet's surface, their heat causes them to melt downward into the ice, cutting a steepening path toward their own demise. This behavior is the opposite of that of rivers on land, which tend to have the steepest slopes in their headwaters and then level out as they approach the coast. In this area of Greenland virtually all supraglacial rivers eventually encounter some crack in the ice, which begins to steal the river's flow and melt open a moulin. After plummeting down the hole, the river tunnels between the basal bedrock and overlying ice, with pressure gradients forcing it toward the ice sheet edge, where the overlying weight (and thus pressure) of the ice is lower. Upon reaching the edge, the meltwater erupts forth into huge muddy rivers flowing to the sea.

One of these, the Watson River, rages past the town of Kangerlussuaq through a deep bedrock gorge. Its whitewater is so wild that my decade-long effort to measure the river's discharge, taking measurements from a steel bridge spanning the chasm, has been futile. And that is just during normal summertime flow. In July 2012, an extraordinary meltwater flood demolished the bridge when the ice sheet's entire surface briefly thawed for four days.

A melting of such magnitude had never been seen in the modern instrument and satellite era. Up on the ice sheet summit, in an ice core drilling camp where temperatures are never supposed to go above freezing, the fluffy dry snow thawed and crusted. At lower elevations, new supraglacial rivers sprang to life and began thundering across the ice. Melting recently surpassed iceberg calving as the leading process by which the Greenland Ice Sheet—a survivor of the Pleistocene whose water has been frozen up on land for over a thousand centuries—is returning home to the sea.

Peak Water

Through the water cycle, sea levels have risen and fallen in response to changes in the Earth's temperature for millions of years. During cold ice ages, more water accumulates in snow and ice atop mountains and continents than is returned to the oceans by melting and river runoff. The outcome is a net transfer of water from oceans to land, causing ice sheets and mountain glaciers to grow and sea levels to fall. During interglacial periods, the melting of mountain glaciers and ice sheets outpaces their accumulation of new snow, causing ice volumes to shrink and sea levels to rise. Evaporation and precipitation are the primary ways in which water is transferred from the sea to the land, and melting and rivers are the primary conduits through which water returns to the sea.

When global climate cools, land ice grows and sea levels fall. When global climate warms, land ice shrinks and sea levels rise. Oceans and ice have danced in this way throughout the Quaternary period, for more than 2 million years. These cooling and warming cycles last roughly 100,000 years, a pace set by wobbles in the precession and axial tilt of our planet and the eccentricity of its orbit around the Sun (called Milankovitch cycles). A single glacial-interglacial cycle thus lasts more than ten times longer than the time that has passed since our first agricultural civilizations emerged.

Today, Earth's climate is warming once again, but at a pace measured in years, not tens of thousands of years. Every year, temperature records are broken and rebroken. Glaciers are shrinking quickly. Global sea level is rising more than 3 millimeters a year. Biological life is migrating toward higher latitudes and elevations. The area of sea ice floating in the Arctic Ocean has fallen by 40 percent since the late 1970s, when NASA first started using microwave satellites to map it.

The chief cause of these observed changes and many others is the addition of carbon dioxide, methane, and nitrous oxide gases to the atmosphere by industrial activity. While other processes also warm global climate—the brightening of the Sun and periods of quiescent volcanoes, for example—all have been carefully measured, and none can explain the magnitude of the temperature increase happening now. The Earth's Milankovitch cycles, which operate over immense spans of time, have no bearing on the dramatic warming of recent years. Only the rising greenhouse-gas content of the atmosphere, which is carefully measured and unequivocal, can explain the temperature increases we are experiencing.

The role that greenhouse gases play in regulating Earth's

surface temperature is uncontroversial and long understood: They warm the troposphere. In the 1820s, the French mathematician Joseph Fourier noticed that the Earth was far warmer than it ought to be, given its distance from the Sun. The physics of *why*—re-radiation of thermal infrared energy by greenhouse-gas molecules in the atmosphere—has been known since 1896, when a Swedish chemist named Svante Arrhenius worked them out using paper and pen. If not for the greenhouse effect, you wouldn't be reading this book, because the Earth would be a lifeless ball of ice.

After more than four decades of research, serious climate scientists agree that the Earth's current warming trend is caused primarily by anthropogenic emissions of greenhouse gases, chiefly from the burning of fossil carbon and the manufacture of concrete. This is unparalleled consensus for a group of skeptical, competitive individuals whose careers demand that they outsmart each other. Ironically, the most common excuses used to deny this hard-won consensus—natural climate changes from other, geophysical causes that happened in the geological past—were discovered by this same scientific community.

These researchers have long since moved on from debating whether or not anthropogenic climate change is real. Their studies now focus on the pace and likely consequences of the rapid climatic and environmental changes we are causing.

This includes many studies of how climate change affects rivers. From analysis of historical discharge records, we know that total annual flows have changed by 30 percent or more in about one third of the world's monitored rivers since the mid-twentieth century. A clear geographical pattern emerges from these records: Discharges in cold, high-latitude rivers are generally increasing, whereas discharges in warm, mid-latitude rivers are generally decreasing, sometimes by 60 percent or more. These declines are especially evident in historical records from China, Africa,

Mediterranean Europe, the Middle East, Mexico, and Australia, with China's Yellow River an outstanding example. Most of these mid-latitude decreases are due to human water withdrawals, dams, and river diversions, with climate change a secondary factor.

The rising river discharges in high latitudes, however, reflect a strong climatic response. Much of this increase is caused by the Clausius-Clapeyron relation, a basic equation in atmospheric physics that describes how warmer air can hold more water vapor (and hence yield more rainfall). Winter flows are also generally increasing, due to milder winters and possibly greater inflows of groundwater, especially on perennially frozen Arctic and sub-Arctic landscapes.

Rivers fed by mountain glaciers are also swelling, as ancient ice melts and returns to the sea. Roughly half of the world's important glacier-fed rivers have already passed through this transient discharge bulge, called "peak water." The effect is temporary. Shrinking glaciers in Asia's water towers are projected to reduce river flows in the upper Indus, Ganges, Brahmaputra, and Yangtze Rivers from 5 to 20 percent by the middle of this century.

This is especially worrisome for the Indus and Brahmaputra Rivers, which derive about 40 percent of their discharge from Himalayan glaciers. Peak water has already come and gone in the headwaters of the Brahmaputra, and it is expected to occur around 2050 in the Ganges and around 2070 in the Indus River. Then, summer flows in these vital rivers will fall, reducing the number of people they can feed by at least 60 million. The projected discharge declines could be even worse if slight increases in rainfall, which are predicted by most climate models, fail to materialize.

There are also some seasonal changes in river flow that occur as land ice shrinks. Big glaciers sustain rivers even during hot, dry years, because extra water is released when more ice melts. During cool, wet years, the glaciers bulk up again. But when glaciers disappear altogether, this valuable buffering effect disappears with them. Today's consumers of glacial meltwater thus face a future of withering river discharge during summers and droughts, when water is most needed for agriculture. Another nasty side effect of declining river discharges is a reduction in the amount of water available to dilute sewage effluent and pollutants. This means that even if pollution levels are held constant, pollutant *concentrations* will increase, worsening the river's water quality.

At shorter, seasonal time scales, mountain snowpacks play a similar role by holding winter precipitation until late spring and summer. When snow stays frozen throughout the winter, downstream river flows are delayed until later, when they are most useful for farming. Where temperatures are too warm to sustain a continuous winter snowpack, this precipitation runs off to the ocean long before planting time. In an effort to mitigate this problem, China, India, the United States, and other countries that rely on snowpack storage for agriculture are building dams to impound some of this winter flow, but it is physically impossible to ever replace the vast storage capacity of mountain snowpacks with reservoirs.

Rising temperatures also remove water from the soil, through increased evaporation and water use by plants. This reduces the amount of water making its way into rivers. This poses a serious problem for the American Southwest, which depends critically on the Colorado River for much of its water supply. Forty million people and virtually all the region's cities rely on this river—most famously Los Angeles and Phoenix, but also smaller cities like Albuquerque and Santa Fe. A combination of declining precipitation and rising temperatures have driven the Colorado River's annual discharges well below their long-term average, creating a structural deficit in its legally apportioned water allocations to seven U.S. states.

Between 2000 and 2014, annual discharges in the Colorado

River averaged nearly 20 percent lower than their long-term (1906–1999) twentieth-century average. About a third to one half of this decline is explained by an unprecedented 0.9°C rise in temperature across its basin, which has reduced snowpack and increased plant water use and soil evaporation. As temperatures rise, annual discharges in the Colorado will continue to decline. Climate model projections conservatively project further losses of at least 20 percent, and perhaps as much as 30 percent, by the middle of this century. By the year 2100, annual flows in the Colorado River could fall to less than half of what they are today.

Our warming climate will affect rivers in other ways besides changing their discharges and timing. As water temperatures rise, warmwater species like carp fare well, but coldwater species, such as trout, do not. River ice becomes increasingly dangerous to cross in winter, and ice breakup floods become less violent in the spring. Although such floods can damage built infrastructure, floodplain ecosystems benefit from the large infusions of sediment and water. On frozen Arctic and sub-Arctic landscapes, thawing ground allows more groundwater to seep into rivers, altering their chemistry and adding ancient, dissolved soil carbon, which is then released into the atmosphere.

One of the most feared effects of climate change on rivers is an increased frequency of extreme floods (except ice breakup floods, as noted). From the Clausius-Clapeyron relation, we know that a warmer atmosphere will generally produce more rainfall. This increases the probability of river flooding, just as higher temperatures increase the probability of heat waves and droughts. However, future precipitation patterns are harder to model than temperature changes, making their prediction a challenging and active area of atmospheric science research.

While historical records document clear, robust trends in total annual river discharge, confirming trends in extreme floods is challenging, in part because very long records are needed to confidently detect statistical trends in rare events. An exception is in the central United States, where statistical studies confirm strong recent increases in the frequency of river floods.

More generally, the picture emerging from climate models is that the frequency of river floods increases non-linearly with global mean temperature. Overall, a 1.5°C increase in global mean temperature is projected to nearly double global human deaths from flooding and nearly triple direct economic damage from flooding. A 2°C temperature increase raises the human toll another 50 percent and doubles the direct economic damage yet again.

However, these globally averaged summaries mask some important geographical contrasts around the globe. For example, flood levels that currently have a return interval of one hundred years (meaning a 1 percent probability in any given year) are projected to become rarer in Mediterranean Europe, Central Asia, and the American Southwest but more common in Southeast Asia, Peninsular India, East Africa, and much of South America. Flood hazards in many places will become more dangerous, therefore, while some will become safer.

While secondary to the effects of diversions, big dams, and water pollution, climate change is pressuring the world's river systems. Long-term reductions in summer flow, changing flood frequency probabilities, and warmer water temperatures challenge human and natural ecosystems alike. These collective stresses make it more important than ever to monitor and understand the world's rivers, and to use their resources wisely. In the next two chapters we will explore how new technologies, sensors, and models will help.

CHAPTER 7

GOING WITH THE FLOW

> The Carmel is a lovely little river. It isn't very long but in its course it has everything a river should have. It rises in the mountains, and tumbles down a while, runs through shallows, is dammed to make a lake, spills over the dam, crackles among round boulders, wanders lazily under sycamores, spills into pools where trout live... The farms of the rich little valley back up to the river and take its water for the orchards and the vegetables. The quail call beside it and the wild doves come whistling in at dusk. Raccoons pace its edges looking for frogs. It's everything a river should be.
>
> —JOHN STEINBECK, *CANNERY ROW*

Nobel laureate John Steinbeck grew up in California's Salinas Valley, a place with a rich history of immigrants, farming, and water conflicts. Today it is mostly known for growing America's lettuce. Many of his novels were inspired by its misfits and landscapes, including the nearby Carmel River, a jewel of a stream that starts in the Santa Lucia Mountains on the Central Coast and flows into the Pacific just south of Carmel-by-the-Sea. In his 1945 novel *Cannery Row,* Steinbeck wrote of the river's lush ferns and crayfish, and of wild foxes and mountain lions creeping along its banks. In his time, the Carmel River flashed and jostled with steelhead, a hardy strain of seagoing rainbow trout.

Steinbeck places his steelhead downstream of a dam, very likely the San Clemente Dam, which was built in 1921 when the future Nobelist was nineteen years old. The towering 106-foot-tall dam provided a critical water supply to the sardine canneries and growing population of the city of Monterey. It also sealed off the upper Carmel River from its steelhead trout.

Like salmon, adult steelhead return from the sea to spawn, laying their eggs in the coarse gravel bottoms of cold, oxygenated inland rivers. A fish ladder (a narrow staircase of stepped pools) built to let fish struggle up and over the San Clemente Dam proved too steep to be effective. Four years after *Cannery Row* was published, a second dam with no fish ladder at all was constructed 6.5 miles farther upstream. The river's spectacular steelhead runs, historically around twenty thousand adult fish per year, began to decline. By the 1960s, fish counts were down more than 90 percent and still falling. In 1997, the steelhead were listed as threatened under the U.S. Endangered Species Act. In 2015, only seven fish were counted at the San Clemente Dam.

Before these dams were built, the Carmel River did what all rivers do—it trundled its load of sediment downstream to the sea. But upon hitting the slack water reservoirs, the river slowed and dropped its burden. Sediment began accumulating behind the dams, and by 2008 the San Clemente reservoir had lost more than 90 percent of its storage capacity. It could no longer serve its original purpose of water storage, and worse, the dam itself had become a safety hazard: The California Division of Safety of Dams determined that it would not withstand a major earthquake or flood. Its owner, California American Water, a utility company, faced a costly retrofitting to save an aging structure that had outlived its purpose.

Fisheries biologists had long decried the dam's blockage of

the river. Rather than retrofit the largely useless structure, the company partnered with the California State Coastal Conservancy and the National Marine Fisheries Service (part of the National Oceanic and Atmospheric Administration) to devise a way to safely remove the dam and restore the Carmel River to a more natural free-flowing state.

The idea was bold and technically challenging. Demolishing the structure would be fairly straightforward, but the abrupt release of nearly a century of accumulated sediment would smother the river in slurry and raise its bed, creating new flood hazards downstream. Excavating and trucking the sediment away was infeasibly expensive. To be cost-effective, more than two-thirds of the sedimentary deposits—about 2.5 million cubic yards—needed to be stabilized in place.

An ingenious engineering solution was devised. The river would be diverted upstream of the dam, bypassing its old reservoir bed and most of the accumulated deposits. A new channel would be blasted through an adjacent ridge, rerouting the river into an approaching tributary creek. The tributary would be widened and regraded into a 2,500-foot-long bypass channel, capable of carrying all of the Carmel River's flow plus its own. Using boulders quarried on site, the bypass was molded into a stepped series of cascades and resting pools, carefully designed to encourage steelhead to pass upstream.

Upon completion of this work in 2015, the San Clemente Dam was dismantled. Almost immediately, steelhead began colonizing the bypass. For the first time since John Steinbeck's teenage years, these endangered fish could freely swim upstream to spawn.

Dambusters Redux

The demolition of the San Clemente Dam reflects a broader movement that is under way in the United States and Europe. As of 2019, nearly sixteen hundred dams had been torn down in the United States alone. While this number is small relative to the roughly eighty thousand (including many small structures) still standing in the country, the trend is gaining momentum. About 70 percent of these removals have occurred since 1999, and old dams are now commonly demolished in the upper Midwest, Northeast, and West Coast states. California, Oregon, and Michigan have all removed more than 5 percent of their dams; Wisconsin has removed more than 10 percent.

Most dam removals are initially motivated by economic concerns, not environmental ones. Sediment infilling and aging concrete have brought many early-twentieth-century structures to the end of their useful life. They tend to be in poor condition and no longer serve the original purpose for which they were built. Unsafe structures burden their owners with liability risk, and hydropower dams in the United States also require periodic relicensing by the Federal Energy Regulatory Commission (FERC), which often involves costly upgrades. In the Pacific Northwest, fast-growing wind- and solar-power capacities are lowering renewable electricity prices, further undercutting the economics of hydropower dams. For economic reasons alone, demolition is therefore an increasingly attractive option for dam owners, especially if there are partners ready to help pay for it. Put simply, tearing out an old dam is often good for the owner's bottom line.

Also quickening the pace of dam demolitions is a growing body of scientific work showing just how effective they are for river restoration. With proper sediment management, removing old dams is less environmentally traumatic in the short term than previously feared. Taking dams down piecemeal over several years tempers the deluge of impounded sediment and associated flooding risks downstream. Within a few short years, as the downstream channel reconnects with its old sediment sources far upstream, the river resumes its old job of carrying its sediment to the sea.

Nearly all of the dams taken down so far have been relatively small structures, mostly under 10 meters tall. But that is changing as engineers and scientists grow more confident. In 2014 two very tall dams, which for decades had obstructed the passage of Chinook (the largest species of Pacific salmon), were demolished on the Elwha River in Washington State. After three years of gradual deconstruction, the final chunks of the 210-foot-tall Glines Canyon Dam (see color plate) and the 108-foot-tall Elwha River Dam were removed.

The recovery was swift. Within a few years, the Elwha River had restored its old channel form and flushed out more than 10 million cubic meters of accumulated sediment, raising the seafloor near its mouth by 10 meters. Because these dams were situated close to the coast, their removal reestablished access to vast areas of high-quality upstream habitat for migratory fish species. Within days of the final blast, Chinook were seen swimming upstream past the old abutments, heading toward Olympic National Park and their ancestral spawning grounds.

An even grander plan is under way to remove four tall hydropower dams on the Klamath River, a heavily subscribed waterway that whipsaws through Oregon and Northern California before entering the Pacific. Native American tribes have been protesting these obstructive dams since the first one was built in 1918. A major salmon die-off in 2002, together with a looming fifty-year FERC relicensing deadline, pitted fishermen and native peoples against farmers and politicians over the future of these aging structures. Yet a remarkably positive and sustained coalition evolved among the disparate stakeholders, eventually producing a series of three grand compromise agreements involving dozens of government agencies, tribal governments, farming organizations, and environmental NGOs.

The hard-won agreements also won the support of PacifiCorp, the dam's owner and a subsidiary of Warren Buffett's

Berkshire Hathaway holding company (further enticement may have come from regular pilgrimages by protestors to Berkshire Hathaway's annual shareholder meetings in Omaha to clamor for the dams' removal). After several years of negotiating, a deal was struck. Pending FERC approval, demolition may begin as early as 2021.

This trend of removing old dams is also under way in Western Europe, where hundreds of thousands of small dams, weirs, and culverts have fragmented river habitats, sometimes for centuries. At least five thousand of these small structures have now been dismantled in France, Sweden, Finland, Spain, and the United Kingdom, with some larger ones targeted as well. In 2018 a 22-meter-tall dam across Spain's Huebra River was demolished to help restore populations of sarda (a small freshwater fish), otters, and black storks in this important tributary of the Duero, one of the largest rivers on the Iberian Peninsula. France is preparing to demolish two large hydropower dams on the Sélune River in Normandy, as well as the large Vogelgrün Dam built across its Rhine River border with Germany.

According to Dam Removal Europe—a consortium of NGOs that includes the European Rivers Network, the World Fish Migration Foundation, the Rivers Trust, the World Wildlife Fund, and Rewilding Europe—thousands more candidate structures have been identified for demolition in the coming years. Economics, liability concerns, and a piece of EU legislation called the Water Framework Directive mandating "good ecological status" and "good chemical status" in European waterways by 2027 are spurring this exciting environmental trend.

Dam removals do pose some risks. Even the best-laid plans for gradually meting out stockpiled sediment deposits can be overwhelmed by an extreme flood. If the sediments contain toxic pollutants, they can devastate a river's ecology if released downstream. Freshly exposed reservoir bottoms seem to attract invasive plant species, and while fish can recolonize their former habitats quickly, riparian forests take decades to grow back. But on balance, the past decade has taught us that rivers can swiftly recover their old physical characteristics after their shackles are removed. Less than one year after the Carmel River was reopened, it had carted off more than a meter of accumulated sediment. Within two years, the river had lined its bed with fresh spawning gravels all the way to the Pacific Ocean.

Rich, developed countries that frenetically built dams across their rivers in the early and mid-twentieth century are now starting to get rid of them. Meanwhile, as described in Chapter 5, the world's developing economies are building and planning thousands of new ones. They too will be grappling with aging concrete, infilled reservoirs, and lost fisheries one hundred years from now. Damming rivers provides many short-term economic and societal gains, but they come with a long-term price.

Starved for Sediment

Another drawback of trapping a river's sediment load behind a dam is the resulting lack of sediment below it. A common problem created by large reservoir dams is the clear, sediment-starved water that they release.

My father, a geologist who has studied river sedimentation for more than fifty years, recently became embroiled in a legal dispute concerning shrinking wetlands on the Saskatchewan River Delta, a vast, ecologically vibrant complex of inland waterways and marshes near the village of Cumberland House in northern Canada. In 1962, the provincial electrical-utility company SaskPower constructed the E. B. Campbell Hydroelectric Station, a large dam and reservoir, across the Saskatchewan River about 60 kilometers upstream from the village. The dam became a significant source of hydropower for the province. But it also caused ecological disaster, in part because it impounded floods that used to regularly spill out across the floodplain and Saskatchewan River Delta, replenishing their abundant marshes with water.

Since the dam was built, the wetlands have been drying up, to the dismay of migratory birds seeking nesting sites and the indigenous Cree and Métis peoples, who for centuries have hunted and trapped the region's abundant wildlife. Reduced populations of ducks, moose, and muskrats present serious hardship for people who rely on them for food, supplemental income, and motivation for youngsters to get outside rather than staying indoors and online.

The local residents quite understandably attribute the continuing disappearance of marshes to the dam's capture of river floods. The problem could be mitigated, they argue, simply by forcing SaskPower to release artificial floods from time to time. But my father's decades of field surveys reveal that the problem is even worse than it seems and is quite irreversible, because of the sediment-starved water being released by the E. B. Campbell Dam.

When a river enters the still water of a reservoir it drops most of its sediment, which settles out on the reservoir floor. As a result, the water released in the river below a dam is stripped clean. The river then cannibalizes sediment from its own channel, deepening and widening it. The eroded, enlarged channel thus becomes less likely to overtop its banks during high-water periods, leaving the surrounding marshes high and dry even during discharges that would have inundated them before.

I first visited the Saskatchewan River Delta more than thirty years ago, as a cheap field assistant on my father's field crew. We were drilling out cores of sand and mud from the river's marshy floodplain, to better understand how its long history of flooding had created one of the richest concentrations of wildlife in

Canada. I remember clouds of ferocious mosquitoes descending upon me as I stood thigh-deep in water and willows, struggling to sink a heavy coring rig into the muck. Today, that site is dry. The muskrats and ducks are gone, as are the riverboats of Cree and Métis people stalking them. The Saskatchewan River, trapped in its ever-enlarging trench, barrels mindlessly on, owned by SaskPower and the physics of sediment and water.

Lessening the Damage

The liabilities and environmental damage of river dams are increasingly spurring calls for their removal, but only in a small club of rich, developed countries facing an epidemic of aging structures built in the early and mid-twentieth century. Meanwhile, a new dam-construction boom is under way, with at least 3,700 major new dams proposed or under construction worldwide. As described in Chapters 2 and 5, most of these new projects are in China, Southeast Asia, Central Asia, and Africa, where domestic opposition to them is weak.

From a global perspective, the environmental gains from the relatively small number of dam demolitions does not outweigh the damage of hundreds of new dams being built every year. As we've seen, the last century of river mega-engineering projects caused serious ecological and social damage. Can the creators of new big dam projects avoid similar mistakes in the next?

The answer lies somewhere between no and maybe. When it comes to huge water storage reservoirs, like those impounded by the Three Gorges Dam and the Grand Ethiopian Renaissance Dam, the tolls of human displacement, altered river flows, obstructed fish, degraded water quality, sediment accumulation, and downstream erosion are unavoidable. However, at least three broad ideas are challenging the old paradigm of towering dams with huge reservoirs that swallow up years of a river's flow and most of its sediment: designer flows, sediment pass-through technologies, and run-of-river dams.

So-called designer flows seek to optimize a reservoir's water releases to serve environmental objectives as well as economic ones. Large reservoirs store a river's natural flow, thus disrupting its riparian ecology. Might this problem be mitigated by strategically timing water releases to mimic natural flood cycles, for example, or to flush out invasive species? One recent modeling study of a reservoir dam impounding the San Juan River (a tributary of the Colorado) concluded that increasing water releases in winter could benefit native fish species and suppress invasive species without damaging the reservoir's basic economic objectives. What's not to like?

In fact, designer flow is a controversial idea that gets people worked up. Take, for example, a recent study that graced the cover of the journal *Science.* It explored whether designer flows might mitigate the impact of planned dam projects on Cambodia's critically important Tonlé Sap fishery, a $2-billion-per-year industry that feeds more than 60 million people in Cambodia, China, Laos, Myanmar, Thailand, and Vietnam. Using mathematical modeling, the authors claimed that carefully timed designer flows might not only preserve this vital fishery but could potentially increase the fish harvest nearly fourfold. The paper was savaged in an extraordinary number of written rebuttals taking its methodology and conclusions to task. The authors returned fire, rejecting their opponents' arguments as "incorrect" and "irrelevant" (heated epithets, for scientists).

More generally, opponents of designer flow argue that trying to combat the overwhelmingly negative environmental impacts of large reservoir dams with the dams themselves is a fool's errand. They note that fiddling with timing releases does nothing to address the problems of blocked fish passage, warmer water, degraded water quality, and sediment impoundment so characteristic of large dams. These are sensible arguments, but they do not refute the proposition that optimally timed water releases—for already existing large dams—might provide some environmental benefits downstream. At the time of writing, it's hard to say who is right. Designer flows are a relatively new idea, and the body of published scientific findings is still small. Many more studies are needed before we can confidently deploy designer flows to help mitigate the environmental damage of large river dams.

A second, more promising idea involves designing and managing new dams such that sediment is allowed to pass through them. The hope is that "sediment pass-through technologies" can mitigate channel erosion below dams and its associated ecological and economic problems. Matt Kondolf, a professor at UC Berkeley who specializes in river restoration, has assessed the risks posed by sediment impoundment behind the more than 130 dams currently being planned for the Lower Mekong River and its tributaries. First, he used computer modeling to calculate the trapping and downstream sediment starvation that these proposed dams would introduce.

Incredibly, their construction (as currently planned) would trap 96 percent of the sediment supply to the Mekong River Delta, creating profound sediment-starvation and erosion problems severe enough to threaten the delta's very existence. In a second study, he identified a range of technological solutions that would allow at least some of the trapped sediment to pass through or around these dams. These solutions include bypass channels, sluicing, flushing, and venting of turbidity currents (a dense, swiftly moving avalanche of water and sediment) through the dam structure. Certain river tributaries carrying high loads of sand would be left undammed. Incorporating these ideas into the region's planning would permit more of the Mekong River's sediment to reach its delta, moderating (but not eliminating) a host of serious sediment-starvation problems.

The third technology that could lessen some of the worst environmental impacts of large reservoir dams is called run-of-river design. The overall concept of a run-of-river dam is to shrink the reservoir or even eliminate it altogether. In its purest form, a run-of-river dam simply allows water to pass through turbines in a low dam, with no water impoundment at all. More commonly, flows are temporarily delayed in a small reservoir, but for hours or days rather than months or years. This strategy won't work in drought-prone areas or for purposes of water storage, but it is appealing for hydropower generation in reliably flowing rivers.

Run-of-river dams don't entirely avoid negative environmental impacts, including altered flows, blocked fish passage, and sediment trapping. However, they are less damaging than traditional large-reservoir dams. Because the Lower Mekong River has reliable year-round flow, all of the large structures being planned there are some variation of a run-of-river dam. If they use sediment pass-through strategies, the Lower Mekong of the future will bear little resemblance to the Colorado River, Yangtze River, and other massively impounded rivers today.

Wheels of the Future

The run-of-river concept is an ancient one, tracing back to the earliest simple waterwheels (see Chapter 1), which rotated a millstone to grind flour. The low-head dams that proliferated across streams and rivers of Europe and New England during the Industrial Revolution were run-of-river, designed to sluice a steady flow onto a waterwheel. This old idea is enjoying renewed interest today, for the purpose of generating modest amounts of carbon-free electricity from free-flowing streams and rivers with little to no impoundment of water.

Microhydropower, or installations having a generation capacity of at least 10 kilowatts and up to either 100 or 200 kilowatts (depending on the country), is a time-honored technological niche that has electrified isolated communities for decades. Tens of thousands of microhydropower installations are currently operating in remote areas of China, Nepal, Pakistan, Peru, Sri Lanka, and Vietnam. A typical setup diverts flow from a mountain stream into a tunnel or pipe, called a penstock, that chutes the water down to a small turbine below, generating enough electricity to power one or several houses. Other variants of this idea use an overshot waterwheel or Archimedean screw. Where discharges are reliable and freezing is not an issue, microhydropower offers a stable, local source of electricity with little environmental damage if the amount of water diverted is kept small.

A number of companies are now selling modern incarnations of these old technologies for urban as well as rural applications. They are manufactured from lightweight sheet metal rather than wood, but the basic idea is the same. In the United Kingdom, a company named Mann Power Consulting has installed more than sixty facilities, most of them modernized versions of the Archimedean screw. Many of Britain's numerous built-up rivers and old mill sites have potential to serve as modest providers of renewable energy, and Archimedean screws are especially well suited for retrofitting existing low-head dams. Waterwheels can be fitted in old millraces, sometimes even in the same wheelpit where a historical wooden waterwheel once turned.

In Northumberland, England, a Victorian-era country manor named Cragside was once the world's first home to be lit by hydropower. Nearly 140 years later, this legacy was renewed when its owners installed a 17-meter-long Archimedean screw

The ancient technologies of Archimedean screws and waterwheels are enjoying a renaissance in today's low-carbon energy revolution. Properly designed systems have little environmental downside and can often be retrofitted into older historical sites, such as this one in the United Kingdom. *(Mann Power Hydro Ltd.)*

microhydropower turbine, through which water from the manor's dammed-up lake is diverted down into its outlet stream. The resultant carbon-free electricity will help Cragside to meet its goal of generating half its energy from renewable sources. In Turin, Italy, a larger project has installed a series of eighty modern waterwheels in an old canal, generating enough electricity to power more than two thousand homes.

While timeless in principle, modern Archimedean screws and waterwheels are a still-evolving technology. A British company named Smith Engineering is selling high-efficiency overshot waterwheels made of laser-cut steel. They generate 5 kilowatts from stream discharges as low as 100 liters per second, paying for themselves in about seven years. The waterwheels arrive in a flat pack rather like IKEA furniture, requiring no special tools for assembly. New Energy Corporation, a Canadian company, sells a vertically oriented turbine that is lowered directly into a stream or river, generating anywhere from 5 to 25 kilowatts. The company also supplies various mounting structures (including an anchored raft that can hold the turbine in a river's current with no impoundment of flow), and conversion kits to transform the generated power into ordinary household electricity.

Small Hydro in Big China

Somewhat larger run-of-river installations are called "small hydropower" or SHP, defined as having capacities of up to either 10 or 50 megawatts (again, depending on the country). China, in particular, has embraced SHP more than any other nation, starting decades ago to electrify remote rural villages. In the 1990s, China further expanded SHP as part of its national environmental policy, to discourage peasants from cutting down trees for firewood.

In the 2000s, after electricity had reached nearly every household in China, the country's use of small hydropower technology evolved again. Connecting SHP installations to power transmission grids became an important part of China's strategy for meeting national targets for electricity generation and reducing greenhouse-gas emissions. To incentivize construction of new SHP projects, Beijing ceded to local governments the authority to approve any privately owned hydropower plant having a generation capacity of 50 megawatts or less. Transmission grid companies were compelled to buy electricity from these plants.

A small hydropower building boom ensued as the technology was transformed from a pragmatic solution to electrify remote villages, to an environmental protection tool, to a national supplier of low-carbon energy for China. By 2015, there were more than forty thousand SHP plants in China, consisting mainly of run-of-river penstock and powerhouse installations. Their total installed capacity approached 80 gigawatts, nearly quadruple that of the Three Gorges Dam.

Over the past two decades, China has greatly expanded its use of small hydropower (SHP) technology. By 2015, there were more than forty thousand SHP plants in China, with a total installed capacity approaching 80 gigawatts, nearly quadruple that of the Three Gorges Dam. This example, along the Nu River, illustrates how discharge from a tributary stream is diverted through a penstock and small powerhouse to generate low-carbon electricity. *(Tyler Harlan)*

China's small hydropower boom has been praised for increasing the country's overall production of renewable energy, but also comes with costs. Because local governments now view SHP as a revenue stream, they have a strong financial incentive to overdevelop small waterways with successive cascades of penstocks and powerhouses. Some streams are literally diverted dry for much of their length, devastating their ecology and harming the livelihoods of local farmers, who can no longer irrigate their fields.

China's experience highlights an important point. Small hydropower installations may reduce greenhouse-gas emissions and be less socially and environmentally damaging than a huge reservoir dam, but they still cause harm if not designed and managed wisely. To address this issue, the nonprofit Low Impact Hydropower Institute (LIHI) hopes to incentivize sustainable installations by certifying qualified micro- and small hydropower projects as having "low impact" on people and ecosystems. For a project to receive LIHI certification, it must have minimal impact on fish passage, recreational users, ecological habitat, threatened or endangered species, and cultural and historical sites, and it must allow free public access at no charge. "Combating climate change should not mean sacrificing ecosystems," observed Shannon Ames, the LIHI executive director. "Low impact certified projects demonstrate that electricity can be generated while still protecting, and often improving, habitats." The idea may be catching on. At the time of writing, nearly 150 projects throughout the United States have received LIHI certification, with dozens of new projects pending.

Innovative companies like Mann Power Consulting and certification programs like LIHI are hoping to nudge the nascent sustainable micro- and small hydropower market toward its potential. According to the European Small Hydropower Association, more than a third of a million possible installation sites are scattered throughout Europe alone. While currently a niche market, properly designed installations are efficient and reliable, and they have little environmental downside. They offer local electricity sources in both rural and urban areas and help contribute to broader renewable-energy goals. They promote interest and education in renewable-energy technologies and can become a focal point of public curiosity when installed in historically significant mills and dams. Isolated mountain communities and Armageddon preppers may first come to mind when thinking about micro- and small hydropower, but urban dwellers seeking to diversify their green energy portfolios can also benefit from this charismatic and low-impact technology.

Snakehead Stew

A soup tureen and two aromatic platters of unrecognizable food steamed on the table in front of me. My companions chatted happily in Khmer and began loading their plates. I peered closer and began to discern forms. On one platter were some tiny freshwater crabs that had been diced, fried, and stewed in spices. On another I saw the bald head and beak of a chicken peeking out from under a chopped pile of its own organs and body parts. Perched on one side was a rubbery brown disk. In the tureen bobbed chunks of flaky white fish and a complete set of cooked fish innards. The smells were tantalizing—chilis, garlic, lemongrass, and *prahok,* a fermented fish paste. I braced myself and tried it all.

It was delicious, even the rubbery puck of boiled chicken blood. The only disappointment was the chicken meat, which was stringy and tough. The hens pecking nervously around us certainly looked rangy and well exercised. During the meal one whizzed past running for its life, hotly pursued by the cook, who had just received another lunch order.

It was a beautiful venue and, like nearly everything in Cambodia, outdoors. We sat cross-legged on the swept wood floor of a raised open-air platform, about ten feet square and shaded by a thatched roof. Scattered around us were other shaded platforms and customers. Before us glistened an emerald pond ringed with lush vegetation. It was spanned by a rickety bamboo footbridge, to which four hand-hewn log dugouts were hitched. The water's glassy surface dimpled and swirled with fish.

The delicate chunks of white meat came from an evil-looking creature called a snakehead. Snakeheads are a class of predatory freshwater fish native to Asia, Malaysia, Indonesia, and tropical Africa. Hours earlier, this particular individual had been yanked from the mud of a drained farm pond and handed to us as a present. It had a muscular, eel-like body with spots like a leopard, a massive head, and a gaping mouth bristling with teeth. It was fearsome to behold, yet, I was assured, delicious. We dropped it off, writhing in a plastic bag, at the restaurant to be cooked for our lunch.

Snakeheads are aggressive hunters of fish, snakes, frogs, birds, other snakeheads, and pretty much anything else they can grab. They survive droughts by burrowing into the mud, as frogs do. They can breathe when out of water and squirm overland to find another stream, river, or pond. I knew horror stories about this fish from North America, where it is a rapidly spreading invasive species due to illegal introductions into the wild. Fish and wildlife managers are petrified by their voracity and resilience. Anglers call them "Frankenfish" and dread their arrival in a favorite fishing hole. I had lectured gravely about snakeheads—and other invasive species like the zebra mussel, brown tree snake, water hyacinth, and Norway rat—to students in my environmental science courses at UCLA.

In 2002, a bewildered fisherman pulled the first reported snakehead in the United States from a pond behind a Maryland strip mall. The discovery prompted a media scrum and the pond's drainage, revealing dozens of snakehead young. They were traced back to a pair of live specimens that someone had purchased in an Asian fish market and released into the pond. Two years later, another snakehead was caught in the Potomac

River, an important tributary of the Chesapeake Bay watershed. By 2018, the fish had spread upstream throughout Maryland and was colonizing nearly three new sub-watersheds per year. At this pace, snakeheads will populate the entire watershed within fifty years, including the Susquehanna, Rappahannock, James, and York Rivers.

It was disorienting to see the same creature that is so reviled in my own country prized as a high-value delicacy in another. In the street markets of Phnom Penh, I saw metal tubs full of live snakeheads that were under fierce negotiation and commanding top dollar from urban consumers. I had known all about the fish's ecological horrors as an invasive species but had no idea it was viewed so favorably at home.

The snakehead we ate was particularly tasty and valuable because of where it came from. It was captured alive at the bottom of a deliberately drained pond but had grown up swimming and feeding in the vast fertile rice fields surrounding it. The fields were dry when I visited in December, but from May through July, during the monsoon rains, they fill with rising floodwaters from surrounding streams, rivers, and Tonlé Sap Lake. During this annual submergence the artificially dug pond lies at the bottom of a vast, shallow wetland less than a meter deep. Adult breeder fish trapped in the pond during the dry season become liberated, swimming long distances through the surrounding rice fields to gorge and lay eggs. Wild fish escaping from the nearby streams do the same. The flooded rice fields explode with planktonic life, providing a rich food base for a short-lived, high-quality freshwater fishery.

Rice-field fish grow fast, reaching eating size in a matter of weeks. Their defecations settle on the bottom, providing free fertilizer to the paddies. When the monsoon rains end, the floodwaters retreat back into the streams, rivers, and artificially deepened catch ponds dug into the fields. The fish have no choice but to follow, and many get trapped in the ponds. By fall, the fields are dry, the rice is tall, and the ponds are packed with catfish, snakeheads, and other marketable fish, naturally funneled from the surrounding areas.

By merging natural seasonal flooding with low-intensity aquaculture, rice fields can become highly productive fisheries. The main requirement is digging some deep, strategically positioned ponds to trap and hold fish through the dry season. Shallow ditches radiating out into the fields maximize the ability of fish to disperse when water levels rise and be captured when levels fall. Some ponds are drained or netted, providing food and cash to the farmers. Others are protected from fishing, to maintain some brood stock. These sanctuaries, called "community fish refuges," serve to repopulate the fields with a new generation of fish the following year.

Had our big snakehead swum into the scenic green community fish refuge next to the restaurant, it would have lived another year. Instead, it got caught in a harvest pond, which was drained. The water was pumped out into the surrounding dry rice fields to irrigate them, and two hundred kilograms of fish were plucked from the mud (see color plate of drained catch pond), dropped into live wells, and hauled off to market to be sold for $1.50 to $2.00 per kilogram. Due to its life on a rice field, our large snakehead, with its unusually clean taste and minimal exposure to chemicals, would have brought $10 in Phnom Penh, roughly twice the price of one raised on a conventional fish farm.

This hybrid form of aquaculture, in which natural and artificial processes collaborate, is called rice field fisheries (RFF). RFF is an increasingly popular form of sustainable fishery being promoted by governmental and nongovernmental organizations in Southeast Asia. I visited Cambodia to visit some RFF projects of WorldFish, an international, nonprofit research NGO that seeks to reduce developing world hunger and poverty through sustainable fishing.

WorldFish organizers travel to rural villages to educate farmers about how they can grow protein alongside rice in their fields. Since most villages lack the financial resources to design and build an RFF system, the NGO approaches donors and governments for grants to fund them. The cost is quite low—with necessary excavations typically costing a few tens of thousands of dollars—and mostly up front, making it attractive to foundations and donors.

One RFF project I visited was constructed in 2015 for the village of Korn Thnot, near Tonlé Sap Lake. I was struck by how much its single, 200-by-500-foot community fish refuge pond, together with some surrounding catch ponds, was helping the villagers. As people cast nets into the catch ponds for their dinner (see color plate), a Cham elder explained how not only his own village but also surrounding villages were benefiting from the annual dispersal of fish into their paddies.

This idea of pairing aquaculture with rice farming is an ancient one. In China it was once common to rotate carp into flooded paddies, and other variants of rice-field aquaculture have been developed and forgotten through the ages. Today, the concept is attracting new interest because it offers a straightforward way to improve the nutritional and financial resilience of the rural poor. More than a billion people in developing countries derive most of their animal protein from fish, and a quarter-million depend on fishing and aquaculture for their livelihoods. Through education, fund-raising, and with heavy earth-moving equipment, WorldFish and other sustainable-aquaculture NGOs are cleverly pairing seasonal flood cycles with agriculture to help some of the world's most impoverished people.

State-of-the-Art Salmon

Nine months before visiting Cambodia, I stood on the deck of a highly mechanized, computerized salmon farm in a fjord in northern Norway.

It was a self-contained steel ecosystem. A central platform radiated long floating black feeding tubes out to ten huge floating rings, ordered neatly in two rows. Suspended under each ring was a deep net enclosure, inside which up to 200,000 fat silvery salmon perpetually circled and ate. They had been transported here after being hatched on land, in a special facility that replicates the natural flow and gravels of a cold freshwater river, bringing full circle Norway's artificial reproduction of the life cycle of the Atlantic salmon. The hoses rattled and shook with oily pellets of brown fish food, automatically meted out by algorithm from huge sanitized steel hoppers bolted to the central platform. Upstairs, a comfortable captain's bridge looked out upon the ring-shaped pens. A wraparound console was lit with glowing LCD screens, monitoring each enclosure's food consumption, temperature, pH, and other water-quality variables. A cheerful Norwegian fish farmer, a mug of fresh coffee steaming in her hand, swiveled around in a padded office chair, explaining all the software. The technology couldn't have differed more from the muddy aquaculture ponds in Cambodia, but in both places the farmer's goal was the same: to grow as much fish meat as possible.

Salmon farming has a checkered environmental history due to infestations of parasitic sea lice, excessive use of antibiotics and pesticides, and food and fecal matter littering the ocean floor. Farmed salmon commonly escape, competing with and sometimes outnumbering wild salmon. Their food pellets are made primarily of ground up fishmeal and oil, meaning other

Salmon aquaculture is flourishing in Norway, where smolts are hatched on land and then transferred to offshore growing farms like this state-of-the-art facility near Bodø. A mechanized central platform (foreground) distributes food pellets to floating ring-shaped enclosures, each containing computerized water-quality sensors and up to 200,000 salmon. *(Laurence C. Smith)*

fish are consumed to grow salmon. But in recent years Norway has made significant progress with these problems, including sharp reductions of antibiotic use, stricter monitoring and quarantining of sea lice outbreaks, and developing alternate feed sources. Salmon farming is a rapidly growing industry that now exceeds $8 billion USD per year in Norway alone, making the small Scandinavian country the world's largest producer of farmed Atlantic salmon. These fish are sold globally and have supplanted wild salmon as one of Norway's most important exports of seafood.

The fish farming trend is global. Already, aquaculture is a $200+ billion industry producing more than 100 million metric tons of food annually, and it is growing faster than any other food production sector. In 2016, the world farmed $139 billion in fish, $57 billion in crustaceans, and $29 billion in mollusks. The five biggest producers are China, India, Bangladesh, Myanmar, and Cambodia. The most commonly raised fish are various carps, tilapia, catfish, and salmon; for crustaceans, shrimps and prawns.

Between 2000 and 2019 farmed fish production nearly tripled, from less than 20 million metric tons annually to almost 60 million. That is a big number, but it still averages to just 7 kilograms of farmed fish per human per year, barely a week's worth of food. With so much upside potential, the fast growth of aquaculture in recent years is likely just the beginning.

In a world that will hold nearly 10 billion people by 2050, this global trend seems both inevitable and desirable. Demand for protein is accelerating, and we possess a ruthless capacity to overharvest seafood, the last bushmeat still pursued at a commercial scale on Earth. Done right, farmed fish, crustaceans, mollusks, and perhaps even algae one day can and should comprise a growing share of the food we eat.

Accidental Aquaculture

One of Germany's largest and most beloved inner-city parks is Tiergarten, a 520-acre oasis of gardens, ponds, and streams not far from Checkpoint Charlie in the heart of Berlin. In summer,

Berliners flock to Tiergarten to relax and enjoy a little nature. But recently, many have been startled by the sight of goggle-eyed crustaceans creeping out of the water and waving their fierce pincers at the walking paths. The park is experiencing a population explosion of *Procambarus clarkii*, the red swamp crayfish, an invasive species native to the streams and marshes of Louisiana. The Tiergarten population most likely originated from the pet trade, when a bored aquarium owner dumped some into one of the park's many bodies of water.

Variously called Louisiana crawfish, swamp crawfish, crawdads, or mudbugs, *Procambarus clarkii* carries a disease that is deadly to Europe's native crayfish species and wreaks havoc on aquatic ecosystems more generally due to its predation, high population density, and aggressive burrowing activities. Unlike most other crayfish, it digs underground to survive dry spells and crawls overland for long distances to find new habitats.

Elsewhere, this unwelcome invader is a valued food. Red swamp crayfish look and taste like miniature lobsters and are delicious boiled, added to spicy jambalaya, or prepared in a host of other ways. The state of Louisiana has more than sixteen hundred crayfish farmers raising them in aquaculture ponds and nearly a thousand commercial fishers catching them along the Atchafalaya River. In total, Louisiana's crayfish industry adds some 150 million pounds of food and more than $300 million annually to the state's economy.

In Germany, officials are granting special commercial fishing licenses and promoting the crayfish's edibility to local restaurants, in hopes that Berliners might eat their way out of the problem. However, shelling crayfish is time consuming, and there is no tradition of eating mudbugs in the country. The strategy is not exactly taking off, and it is likely that this invasive species will persist and expand its range.

Still, the idea of eating invasive crayfish has precedent. A century ago, someone in Nanjing got the idea to import some *Procambarus clarkii* for frog food. The species soon invaded the area's rice paddies and was considered a noxious pest for decades. But in the 1990s, local chefs began cooking them with spices, and the pest became a sought-after delicacy. Today, crayfish aquaculture is a $2 billion per year industry in China, and the country's exports of cheap frozen crayfish tails put Louisiana's producers at a serious competitive disadvantage.

With precedents like this, it is fascinating to contemplate whether developing new food markets might help to fight the spread of invasive species threatening riparian ecosystems around the world. Take, for example, the serious problem of Asian carp, which are spreading into the Mississippi River watershed and are on the verge of penetrating the Great Lakes. Originally imported from China in the 1960s and '70s to control algae in ponds, they escaped into the Mississippi during big floods and have since expanded into the Missouri and Illinois River systems as well.

The term *Asian carp* actually comprises several species that have been domesticated in Asia for centuries, notably bighead carp, silver carp, black carp, and grass carp. They grow prodigiously in both size and number, seriously outcompeting native fish, altering water quality, and injuring boaters—when panicked, the carp leap high into the air. Several fatalities have been blamed on leaping Asian carp, which can grow to 100 pounds or more. Their phenomenal leaping ability also allows them to sail over low-head dams, locks, and other structures that would otherwise block their passage.

At the time of writing, a series of electric barriers installed in Chicago's canals by the U.S. Army Corp of Engineers is the last remaining deterrent stopping Asian carp from breaching Lake Michigan. If that happens, they could wreak unimaginable damage to the $7 billion per year fisheries industry and Great Lakes' native ecology and begin colonizing the many rivers that flow into them. The threat is so serious that the Michigan Department of Natural Resources recently announced a $1 million prize to anyone who can devise a realistic plan to prevent the entry of Asian carp into the Great Lakes.

Might growing a commercial fishery for Asian carp help to control this invasion of the American heartland? They are a bony fish but have a mild white meat that is quite good when ground into patties or fried. When grilled, it holds together like swordfish. The Southern Illinois University of Carbondale is studying the use of Asian carp for fertilizer and livestock and aquaculture feed. The University of Illinois is experimenting with serving Asian carp in its dining halls. In Kentucky, a young company called FIn Gourmet Foods has patented a way to remove the fine bones from Asian carp and is processing them into boneless fillets, fish cakes, fish burgers, and surimi (a fish paste, commonly called imitation crab meat, or krab, in the United States). For just $71, I was able to order frozen Asian carp burgers, fish cakes, and sesame ginger marinated fillets directly from the company's website. I found the cakes to be a bit rubbery, but the fillets were delicious, with a firm, flaky texture rather like that of whitefish.

Silverfin Group, a private venture, believes that a sizable domestic market can be grown. As their name implies, the group proposes rebranding Asian carp as "Silverfin," and in 2018 it partnered with Sysco Food Services to distribute Silverfin products across the country. "By promoting commercial harvest of Asian carp," goes the group's mission statement, "we will reduce their population to minimize their threat so they can coexist with native species, minimize water sports accidents caused by jumping carp, create much needed jobs, revitalize commercial fisheries, stimulate local economies, and provide a clean and healthy fish product for consumers." However, even with an image makeover, creating a helpful domestic consumer market for this fish is a daunting challenge. Most Americans find carp unappealing, and a huge market is needed. An adult female can spawn more than 1 million eggs annually, with most hatching and surviving. With fish yields estimated at 35 tons of carp per river mile, and catch rates approaching three thousand pounds per hour, millions of Americans would need to discover a taste for Silverfin for the country to eat its way out of this invasive species problem.

A more realistic market is China, where Asian carp is already popular and consumers perceive wild-caught American fish as safer to eat than domestically farmed fish. The chief barriers to growing this international market are the low price of farmed carp and a consumer preference for live fish. But despite these challenges, a new export fishery may have begun in America's heartland.

The State of Illinois has introduced several business initiatives intended to spur a viable Asian carp fishery, including a $2 million grant to the Big River Fish Corporation, a company seeking to export American Asian carp to China. In Kentucky, an entrepreneurial woman named Angie Yu relocated from Los Angeles to found Two Rivers Fisheries, also eyeing this export market. Her company hires commercial fishermen to catch the carp, which are then processed into patties, tails, sausages, ribs, and dumplings and flash frozen for shipment. Two Rivers Fisheries processed a half-million pounds in 2014 and 2 million in 2018. Their production is growing rapidly, with a target of 10 million pounds by 2021 and a whopping 20 million by 2024. Ninety percent of the Asian carp this company produces is exported overseas.

Accidental aquaculture is at best a partial solution to the problem of invasive species spreading through the world's waterways. Credible commercial markets may exist for Louisiana crayfish and Asian carp, but not for Great Lakes zebra mussels or Potomac River snakeheads. In the case of Asian carp, a reliable supply chain must be built from scratch, including ice houses, processing facilities, and large numbers of commercial fishermen that must be trained to catch an unfamiliar species. If people become comfortable eating Asian carp, and their commercial value becomes established, the likelihood of further illegal transplants only increases.

Furthermore, growing a new industry around an invasive species is no way to extirpate it, since a successful business model requires maintaining or even growing its supply. Should the budding Asian carp industry take off in the Mississippi River valley, I expect it will form a trade group and hire lobbyists to protect its business, just like any other industry. Commercial harvesting could help to control Asian carp populations in America or Louisiana crayfish in Germany, but is highly unlikely to eradicate them.

New Roles for Old Rivers

So far, this chapter has explored some promising new ideas that could benefit rivers by restoring them to a more natural state—through dam removals, new dam designs that impound less water and sediment, micro- and small hydropower, sustainable aquaculture, and eating invasive species. Physical river restoration (through removal of concrete linings and flood control structures, for example) is a rapidly growing subfield in civil engineering that will be discussed with regard to the Los Angeles River in Chapter 9. Together with pollution controls (see Chapter 6), all of these have potential to reverse humanity's long history of degrading the environmental health of rivers.

These ideas deliver good societal benefits, including recreational areas, sustainable hydropower, and fish protein. But they are modest in scale and focus largely on mitigating mistakes of the past. Might some new big innovations surrounding rivers be coming? Or are sizable social and economic gains limited solely to huge reservoir dams, an early-twentieth-century idea dating all the way back to the Hoover Dam, Grand Coulee Dam, and Tennessee Valley Authority?

The Three-Billion-Dollar Battery

One of the more provocative big new ideas currently under discussion is the possibility of turning the Colorado River into a giant contraption for renewable-energy storage.

As the cost of generating renewable wind and solar electricity continues to fall, the main deterrent preventing utility companies from adopting these technologies becomes their intermittency, not price. When the sun sets and winds falter, these sources generate too little electricity. When gales blow and skies are sunny, they generate too much. If the surpluses can't be sold or given away, they must be shut down or risk overwhelming the grid. Neither utility companies nor consumers can tolerate such volatility, an important reason why coal- and natural gas–fired power plants are an essential part of our energy mix today.

A nationally distributed high-voltage transmission grid would greatly help to smooth out intermittency problems, but in the United States it would be expensive and politically difficult to create. Also, energy is lost when transmitted over long distances. Therefore, part of the solution requires finding local, cost-effective ways to store surplus electricity so it can be returned to the grid later when needed. Many studies and pilot programs are exploring a range of exciting technologies, ranging from giant battery warehouses to pumping compressed air into underground caves. One idea involves hoisting concrete blocks when electricity is abundant then dropping them to generate electricity later. To date, most energy-storage schemes are either too expensive or have yet to be demonstrated at large scale. A distinct exception is pumped water storage, a time-honored technology that has been used for decades to smooth out variations in hydroelectric power supplies.

Pumped water storage is conceptually straightforward and typically uses existing hydropower dams. As surplus electricity becomes available, water is pumped from below the dam's powerhouse back up into the reservoir, where the water is stored. When electricity is needed, the water is re-released through the dam's turbines, generating electricity. While some energy is lost, the recovery efficiency is around 70 to 80 percent. Pumped water storage is not a global solution because it is limited to the same types of physical sites needed to build reservoirs, but it is a popular and effective technology where conditions allow. It is already in wide use in China, Canada, the United States, the Russian Federation, and other heavy producers of river hydropower.

The Hoover Dam impounds the Colorado River in Lake Mead, a huge artificial reservoir near Las Vegas. As described in Chapter 6, declining river discharge and a structural deficit in water withdrawals have driven Lake Mead's water levels to extraordinarily low levels; at the time of writing it is operating at just 20 percent of its power-generation capacity. But in 2018, the Los Angeles Department of Water and Power (LADWP) announced that a major engineering study was under way to explore the possibility of making the reservoir an integral part of a new $3 billion pumped water storage system.

What makes the project so unusual, aside from its price tag, is that the energy would not come from conventional energy sources. Instead, the proposal is to store energy generated solely by wind and solar farms, at a still-undetermined location around 20 miles downstream of the Hoover Dam. During times of surplus renewable electricity generation, water would be pumped back up behind the dam through a newly built pipeline. When electricity is needed, the water would be released through the dam's turbines.

The idea is groundbreaking because of the sheer scale of the pumped water storage that would be provided, and because all of the energy would come from carbon-free renewable sources. As envisioned, bringing this much energy storage online would sufficiently mitigate the intermittency problem to allow California to meet its goal of running on 50 percent renewable energy by the year 2030.

The nominal timeline for this grand proposal to turn the Colorado River into a $3 billion battery is 2028, but many engineering, economic, and political hurdles remain. Because the Hoover Dam is on federal land, the U.S. Department of the Interior would have to sign off. The National Park Service would conduct reviews and hearings to assess the proposal's environmental and cultural impacts. Downstream, the residents of Bullhead City, Arizona, worry about disruptions to recreational boating in Lake Mohave, another artificial reservoir on the Colorado River that already experiences significant water level fluctuations.

While pumped water storage is not new, this proposal's massive scale and singular focus on storing renewable wind and solar energy most certainly is. Should the project come to pass, it will mark one of the most unusual new ideas for human river use in the early twenty-first century.

Empty Your Bowl

A second big idea is so obvious that it's easy to overlook. As detailed in Chapter 4, river flooding poses an ever-present threat to humankind. Our growing cities have intensified this threat by developing risky, low-lying areas that were prudently avoided by previous generations. On river deltas this threat is further exacerbated by rising global sea level (caused by climate change), sinking land (caused by groundwater pumping), and increased coastal erosion (caused by trapping of river sediment behind dams). Faced with these pressures, any new technology that reduces flooding represents a direct and immediate benefit to society.

One of the world's most endangered cities is tackling this head-on through new flood defenses and pumping technology.

New Orleans, Louisiana, sits in a bowl. More than a million people—nearly a third of the state's population—live in and around a squiggly strip of low-lying land bounded by the Mississippi River to the south and Lake Pontchartrain to the north. Most of the city is at lower elevation than both of these water bodies. River barges slide past, often higher than building foundations. The soil is so saturated that the dead cannot be buried below ground. They are interred in mausoleums, raised above the surface.

The coveted high ground in New Orleans follows the riverbanks themselves, on natural levees that accrued over centuries when the Mississippi could still freely overtop its banks during floods. When rapidly moving floodwater overtops a riverbank, it slows down, dropping its heaviest particles of sediment and piling up a sinuous ridge along the bank called a natural levee. For millennia, settlers of flat, swampy deltas and lowland river valleys have prized this high ground. The founders of New Orleans built the historic French Quarter atop one. In terms of flood danger, one of the city's most raucous districts is actually one of the safest.

Put simply, there is no natural drainage in New Orleans. Almost all water that descends on the city—whether from a Mississippi flood, a hurricane surge from Lake Pontchartrain, or just a little rainstorm—must be pumped out. And that is exactly what the city is doing to survive.

"The pump operators are our first responders," explained Angela DeSoto, a project manager for the Southeast Louisiana Urban Flood Control Project (SELA), an ambitious new pumping scheme. "Without them, we flood." I was with her and three others touring the "safe room" of a new SELA pumping station in Jefferson Parish, about twenty minutes from downtown New Orleans. The station is built strong enough to withstand a category-5 hurricane. It has steel bunk beds, off-grid power generators, food and water stockpiles, and an escape hatch through the ceiling. In the event of flood disaster, the pump operators are forbidden to leave.

SELA is an ambitious new partnership between the U.S. Army Corps of Engineers, the State of Louisiana, and local parish authorities. To date, it has secured $2.7 billion to build powerful pumping stations, box canals under the streets, and a "Pump to the River" that carries water off the city streets and into the Mississippi River. Its three massive pipes emerge from the ground

An important component of Louisiana's ambitious new Southeast Louisiana Urban Flood Control Project (SELA) is a "Pump to the River," which lifts runoff from the streets of greater New Orleans into the Mississippi River. Photographed here are some of the system's outflow pipes rising up and over the levee into the river—see person for scale. *(Laurence C. Smith)*

and rise up and over the levee. Through this design, SELA has effectively turned the Mississippi into a massive storm drain to help empty out the city.

SELA will constitute a third rail of flood defenses for the greater New Orleans metropolitan area. The first two are the Hurricane & Storm Damage Risk Reduction System (HSDRRS) and the Mississippi River & Tributaries Project (MR&T). The HSDRRS is a 133-mile-long defensive perimeter of floodwalls, levees, gates, and pumping stations that was begun in 2005, after hurricanes Katrina and Rita. The MR&T dates to the national 1928 Flood Control Act (yet another outcome of the massive 1927 Mississippi River flood described in Chapter 4). This sweeping bill mandated the U.S. Army Corps of Engineers with an eternal grapple with the Mississippi River. Today, the MR&T consists of more than 4,000 miles of artificial levees, 2 million acres of floodways and backwaters, reservoirs and pumping stations, and numerous modifications to the Mississippi River channel itself.

The fight rolls on to save New Orleans, a great American city that celebrated its three hundredth birthday in 2018. At the time of writing, thirty SELA projects have been completed in Jefferson Parish, and twenty more are either finished or under way in Orleans Parish. Once completed, the SELA partners hope to secure an additional $2.3 billion to build similar infrastructure in the surrounding parishes. If successful, some $5 billion dollars will be invested in the Southeast Louisiana Urban Flood Control Project, to help protect one of the world's most flood-endangered metropolises.

A Dark Desert Highway

Two thousand miles west of New Orleans, another great American city is seeking technological solutions to its river problems.

Los Angeles, as you may recall from Chapter 5, is able to exist in its modern form only by tapping faraway rivers. To the consternation of Northern Californians and residents of six other western states, their Sacramento, Owens, and Colorado Rivers are all used for the well-being of Southern Californians. The water is diverted south and west through the California, Los Angeles, and Colorado River Aqueducts. Together with smaller, local rivers (especially the Santa Ana and San Gabriel Rivers) and groundwater aquifers, these three artifical rivers supply water to 19 million residents of Los Angeles, Orange, Riverside, San Bernardino, San Diego, and Ventura Counties.

Whether they realize it or not, they are customers of the Metropolitan Water District, the wholesaler and overseer of most of Southern California's water supply. This gargantuan cooperative of cities and municipal water districts, first presented in Chapter 5, formed in 1928 (the same year the U.S. Congress passed the Flood Control Act). Metropolitan's original founders included the cities of Anaheim, Beverly Hills, Burbank, Colton, Glendale, Los Angeles, Pasadena, San Bernardino, San Marino, Santa Ana, and Santa Monica. Today, its twenty-six members include fifteen cities and the central water authorities of the Inland Empire, Orange, and San Diego Counties. Its physical facilities include nine reservoirs, sixteen hydropower plants, five water treatment plants, and nearly a thousand miles of pipelines, tunnels, and canals.

~

I first met Jeff Kightlinger, Metropolitan's leader and general manager, in the executive suite of the organization's grand headquarters in downtown Los Angeles. Straddling the building's entryway are two huge ceramic tile murals. One depicts a concrete aqueduct sliding through a desert toward distant skyscrapers, sowing a cornucopia of vegetables along the way; the other a towering dam spewing five frothing jets of blue water. Inside, on the lobby floor, appears a motif featuring an eagle, golden bear, and two burly, shirtless workmen aiming a gushing pipeline at the viewer. The Metropolitan Water District owns up to what it does.

Over the course of an hour, Kightlinger described many challenges that Metropolitan has overcome during the last century and those it will have to overcome in the next. Metropolitan was a driving force for the State Water Project, California's grand north-to-south river diversion scheme described in Chapter 5. It built the Colorado River Aqueduct, which carries Rocky Mountain water 242 miles from the Arizona border. It has the power to negotiate directly with other states and Mexico, and recently wrote a water-sharing agreement with Arizona and Nevada for the Colorado River. Each year, its chemistry labs perform nearly 250,000 water-quality tests for constituents throughout its water-distribution system. With nearly two thousand permanent employees and an annual budget approaching $2 billion, it is an economic, as well as existential, force in the region. Were the German historian Karl Wittfogel (see Chapter 1) still alive today, I believe he might declare Los Angeles a modern hydraulic society, and the Metropolitan Water District its most essential and implacable bureaucracy.

Looking ahead, Kightlinger is now steering Metropolitan in a radical new direction: endlessly recycling the water that it imports. A personal epiphany struck him in 2015, during one of the worst water shortages in Southern California's history. "If not for the State Water Project," he said grimly, "L.A. would have been crushed by those droughts." The crisis sharpened his focus on something he and others had realized for some time—that Metropolitan would once again have to find more water for the growing region. And with no faraway natural rivers left to tap, some unnatural ones would have to be found.

Historically, Metropolitan had focused solely on water supply infrastructure, not regional water planning. But after an earlier drought in 1991, Metropolitan began releasing forward-looking documents every five years. Hiding inside these blandly named Integrated Water Resources Plans are a chronicle of new ideas that are giving the giant consortium a new role as the region's foremost water planner.

First, Metropolitan consolidated its power over its fractious members by imposing standardized measurements, metrics, and water-conservation programs. To fund the programs, it levied on its members a universal tax (called a water stewardship fee) of about $80 per acre-foot. The revenue from this tax is used to develop new, locally sustainable water sources. Bringing these new local supplies online is one part of Metropolitan's broader strategy for meeting the region's growing water demand.

To this end, Metropolitan is now turning to treated wastewater as a new source of water supply. In 2017, it announced a new partnership with the Sanitation Districts of Los Angeles County (LACSD) to create the Regional Recycled Water Program, a plan to reuse treated sewage effluent by purifying it and then pumping it underground. For its wastewater source, Metropolitan has targeted a huge LACSD facility in Carson, California, one of the nation's largest sanitation plants, which currently discharges its treated wastewater to the ocean.

In 2018, a pilot demonstration facility was built inside the Carson compound. After collecting data for about a year, the experience will guide construction of a full-scale facility on the site. As currently envisioned, the Regional Recycled Water Program will produce up to 150 million gallons a day of purified drinking water and transport it through sixty miles of new pipelines radiating from Carson to four groundwater aquifers serving 7.2 million people in Los Angeles and Orange Counties.

To put that 150 million gallons a day into perspective, the

Carson water treatment plant currently receives about 400 million gallons a day of raw sewage, the waste stream of 5 million people. This recycling program, therefore, seeks to return more than a third of that in the form of clean, drinkable groundwater. It is equivalent to returning nearly half the flow of the Los Angeles Aqueduct underground.

I attended Metropolitan's groundbreaking ceremony for the Regional Recycled Water Program pilot demonstration facility in Carson. The event was held on site, with a big shade tent and stage set against a sprawling backdrop of pipes, tanks, and digester chambers of the LACSD sewage treatment plant. There were catered fig sandwiches, California fruits, and chilled bottles of purified wastewater to drink. Politicians were milling about; many speeches were made.

One of the best was delivered by Kightlinger, who reminisced about his first job working for a sanitation district and how he always wondered why water-supply and water treatment agencies never talked to each other. "We work really hard to bring water into Southern California," he roared, "what a waste to use it once then ship it outta here!" The audience cheered and clapped and drank more purified sewage water. Two dignitaries walked over to two gold-plated ceremonial faucets and opened them, to symbolize turning on the tap of an important new water source for Southern Californians (see color plate). I could not help thinking of William Mulholland turning on the Los Angeles Aqueduct at a similar ceremony, more than a century before.

Groundwater spreading, the technical name for this method, has a long history in Southern California. However, it has historically been done by capturing water from rivers, not sewage treatment plants. As explained to me by Michael Wehner, the assistant general manager of the Orange County Water District Groundwater Replenishment System (GWRS), Orange County is already running the world's largest groundwater recharge program and will become one of Metropolitan's main customers for the purified water from the Carson plant.

All along the normally dry gravel bed of the Santa Ana River, the GWRS has dug a series of rectangular catch basins. The basins fill up during floods, then the captured water slowly percolates into the ground. Flood water is also diverted into some deep abandoned gravel quarry pits nearby and, again, allowed to soak into the ground. Rather cleverly, Orange County created these pits by purchasing the undeveloped land, then selling gravel until they got the holes they wanted. The proceeds from the gravel sales even repaid the cost of the land.

Today, the Orange County Water District Groundwater Replenishment System is an impressive and award-winning setup that is running below production capacity. Flows in the Santa Ana River have declined because other water districts farther upstream have caught on and are trapping and percolating the river's water too. Orange County is eager to buy Metropolitan's recycled wastewater product and has the necessary spreading basins already in place. Because the water will be percolated into existing aquifers that are already in use for human consumption, rather than piping it directly into the city's water supply, this will help mitigate the psychological yuck factor that has confounded other "toilet-to-tap" water recycling efforts to date.

Treated wastewater is the last untapped source of river water available to Southern California. At the present time, it runs off to the ocean. Not for long. With an estimated construction cost of about $2.7 billion, Metropolitan's Regional Recycled Water Program rivals the LADWP pumped water storage proposal for Nevada's Lake Mead in ambition. If it comes to pass, the original source of much of this new water will be the Owens, Sacramento, and Colorado Rivers. Then—like "you" in the classic Eagles song "Hotel California"—it will never leave Los Angeles again.

CHAPTER 8

A THIRST FOR DATA

I am not a religious man. I am compelled by things I can see and touch. I believe that science is our best tool for exploring our world and universe, even though we are doomed to never fully understand either. Measurements, observations, and numbers provide comfort and solidity for people with a worldview like mine.

That is why I am still so unsettled about an odd thing I saw, with my own disbelieving eyes, on top of the Greenland Ice Sheet on July 19, 2015. Had my colleagues and students not been yelling and gesticulating, I'd have thought I was hallucinating. But the two objects floating swiftly into view were as real as the ice beneath my feet, the crates of gear piled around me, and the bright red Air Greenland helicopter powered down on the ice sheet nearby.

Despite their quickening speed and low profile in the racing supraglacial river, I knew immediately what they were. Each had a white, ring-shaped life preserver, the sort found hanging on the walls of swimming pools and cruise ships, and a bright orange Pelican case fitted into its center hole. I knew that beneath the ring were rigid polycarbonate fins that protruded a few inches down into the water to catch and ride the current. I knew that inside each case was a GPS receiver, a sonar echo sounder, a temperature probe, and an Iridium modem to continuously beam the device's precious measurements up to orbiting telecommunications satellites, before it plummeted into the bowels of the ice sheet.

One floating ring was swiftly gaining on the other, to the whoops and hollers of my field team. Somehow, the trailing device caught up with the leading one at just the moment both passed under a long rope cableway we had slung, with considerable difficulty, across the blue torrent. Like straining racehorses, they passed together under the rope, briefly united, before being swept into the thundering moulin a few hundred meters downstream and shattered into oblivion.

My neck hairs prickled and a strange, alien feeling descended over me. A couple of hours earlier I had dropped those same devices into two separate upstream tributaries, more than 2 kilometers apart, with about thirty minutes' time between each launch. As planned, each had rollicked down its twisting tributary to end up in the river in front of me, and from there to be swept off into the moulin. But the two devices should never have encountered each other. For them to reunite under our rope cableway after riding out two totally separate whitewater trips, in different river channels, traveling at different speeds, getting hung up in whirlpools, and a host of other timing improbabilities, was about as likely as me walking into a casino with a nickel and walking out a millionaire.

A few hours earlier, my colleagues and I had gathered in a semicircle around these same devices just yards from where I was standing, to honor Dr. Alberto Behar, the NASA robotics engineer who had designed and built them. Alberto was supposed to be there with us on the ice sheet, but he was dead. Six months earlier, his experimental single-engine plane had dropped out of the sky onto a busy San Fernando Valley street moments after departing Los Angeles' Van Nuys Airport. Instead of Alberto, it was I who had packed his pioneering devices into the red Air

Greenland helicopter and flown off to launch them in two different, distant headwaters of the supraglacial river rushing before us.

Alberto was a good friend to many of us who were there that day, and his death hit us hard. During our tribute one anguished colleague expressed conviction that he was there with us on the ice sheet, a place he had visited more than any of us. I recall wishing that I felt that way too, but I didn't. But now, after what I had just seen with my own incredulous eyes, I wasn't so sure.

Later, flying low over the ice back to Kangerlussuaq, I insisted to my helicopter pilot the impossibility of what had happened. He shrugged. "Maybe your floats had some help." Maybe so, I replied uneasily. And one day, when I summon up the nerve, I'm going to open up the precious data those sensors transmitted, and run them through a watershed model. Then, I'll learn if what I saw was inexplicable by science—or not.

~

Alberto Behar's life passion, besides his lovely wife and three young children, was building autonomous sensors and remote sensing systems to collect rare scientific data in extreme environments. The instrumented river floats and an autonomous boat (see color plate) he built for my work in Greenland were just two of his many inventions. As a robotics engineer at the NASA Jet Propulsion Laboratory in Pasadena, California, he designed instruments for the Mars Curiosity rover and the Mars Odyssey orbiting spacecraft. As a professor at Arizona State University, he directed the Extreme Environments Robotics and Instrumentation Laboratory, where he and his students devised cunning sensors and imaging devices to study the deep ocean, icy polar lakes, fuming volcano craters, and other insanely challenging field sites.

He was at the forefront of a quiet revolution that is transforming how scientists study and monitor the natural world. A dizzying array of small, cheap sensors are coming online, driven especially by the emerging autonomous vehicles market. Adapted for scientific use, these components can be left outdoors for months or years, mounted on drones, linked wirelessly to cellular networks, or just sent off on suicide missions like our floating devices in Greenland. Imagery and data from new satellite remote sensing technologies are being freely disseminated online by NASA and other national space agencies, and, increasingly, by private companies. Taken collectively, these technologies are rapidly expanding our understanding of what transpires on the surface of the Earth. Alongside numerous other pragmatic uses, they are monitoring our planet's oceans, land, ice, vegetation, human activities—and rivers.

This data explosion could not have arrived at a better time. Our scientific understanding of rivers has many gaps and is still evolving. If you've read this far, then you well know that the world's waterways are under pressure and will be taxed even harder in the future. But how have we been able to recount all these stories so far? Because someone, somewhere, recorded sufficient information for them to be told. And because rivers are so vast, so important, and so intertwined with human and biological life, countless other stories were missed and are now lost to history. In a world of 8 billion people and nearly two hundred countries, with countless river-related challenges for people and ecosystems, we can't afford that. We need machines to attain the science and to help fill the information gaps. We need sensors, satellites, and models.

The River's Purpose

For billions of years, flowing water has battled plate tectonics for dominion over the continents. When plate collisions thicken and raise the Earth's crust, rivers attack and start leveling it back down. Using the water supplied by climate, they convey the eroded material toward the sea. The details of this process determine much of what landscapes look like and how rivers behave.

If you've ever ridden a car high up into rugged mountains, your road probably followed a river valley. As your vehicle climbed higher in elevation, the valley below likely narrowed and its river came into view. You probably gazed down at a rocky, boulder-strewn gravel plain and picturesque rushing cataract. No farm fields along those banks—the cobblestone corridor looked rubbly and raw, more suited for marmots than plows.

As your car continued to climb, the gravel-filled valley began dropping away beneath you. It was cutting down into the mountain, almost as if in defiance of the rising peaks. The river was imposing its own, gentler slope onto the mountainous terrain.

Eventually the rubbly valley probably narrowed then disappeared. Your road strained higher, around knuckle-whitening hairpin turns, likely passing some pretty waterfalls and jewel-like lakes where mountain glaciers once perched. When you finally reached the pass, perhaps you took a few hurried photos in the cold wind before driving on. Then the road began dropping again, and a new boulder-strewn valley came into view before you. Mirror-like, the vista repeated as you descended down the other side of the mountain into another gravel-filled river valley below.

The pass, where you may have snapped some photographs, is the topographic divide, the front line in the two rivers' battle for supremacy against the mountains and each other. Fueled by rain, snow, and gravity, each is carving down a mountain flank. Their eroding headwaters cut uphill like knives toward each other. The rivers are pulverizing the mountains and carrying them off to the sea. One day, when the underlying tectonic forces tire of raising the range, the rivers will win. It will take tens or hundreds of millions of years, but they will win.

Below the lakes and waterfalls, where the valleys widened and became buried in gravel, the rivers were asserting themselves. This is typically where mountain pass roads merge with river valleys and gratefully follow their gentler slopes down onto the flat plains below. All rivers, of course, are constrained by their topographic surroundings and underlying geology. But within these constraints, rivers operate by engineering their local environment. The reason seems almost purposeful: to carry off the eroding mountains, particle by particle, as energy-efficiently as possible given the amount of sediment and water available.

The river achieves this by adjusting the slope of its own bed. Water observed flowing over waterfalls and clean, hard bedrock is slowly cutting its way downward. However, once a stream or river picks up enough material to also *deposit* sediments, it begins to take charge of its immediate environment. Here, the river lays down material; there, it removes some—all with the goal of achieving just the right slope necessary to move its assortment of cobbles, gravel, and sand downstream.

Where the rock fragments sliding into the river are large—along our rubbly mountain valley, for example—the river steepens its slope, so as to increase its flow velocity and tractive force to roll those big rocks along. As they bounce downstream, they break progressively into smaller cobbles, then gravel, then sand, then silt. In concert with this sequence, the river gradually lessens its slope to become gentler, at just the right angle to keep this ever-smaller, lighter, easier-to-move sediment mixture moving downstream. Along with depositing sediments, the river also starts meandering, creating those great lazy, looping horseshoe-shaped bends so ubiquitous along lowland river valleys. Meandering increases a river's length over the same drop in valley elevation, thus decreasing its slope.

The physics of all this is complicated and highly mathematical. It remains an active area of research even after decades of study. It is a subject that greatly fascinated a certain Nobel Prize–winning theoretical physicist named Albert Einstein, but he decided the subject was too difficult and turned to astronomy instead. His son, Hans Albert Einstein, would go on to spend his entire career at the University of California at Berkeley pursuing what his father did not: an improved mathematical understanding of how rivers move their sediment downstream.

The idealized case, which many rivers approach but few achieve, is a sort of equilibrium state called a graded river. A graded river, wrote the American geologist J. Hoover Mackin in a seminal paper, is one "in which, over a period of years, slope is delicately adjusted to provide, with available discharge and the prevailing channel characteristics, just the velocity required for transportation of all of the load supplied from above." The origins of this idea trace back even earlier, to the late nineteenth-century work of G. K. Gilbert and William Morris Davis, the founding co-parents of modern geomorphology (the study of landforms and how they are created).

Graded rivers display what is called a classic concave-up longitudinal elevation profile, meaning that the slope of the riverbed gradually flattens out, like the long handle of a gently curving hockey stick, from its headwaters to mouth. The upper reaches of a graded river have a steep slope with coarse sediments lining its bed. Downstream, the river's slope becomes progressively gentler and its sediment finer. The reasons for this boil down to a balance of forces between a river's gravitational energy and the amount of energy needed to move the sediment and overcome the frictional resistance of its bed.

Many things can and do prevent rivers from achieving this idealized equilibrium state, including tectonic uplift, tributaries, sea-level rise, and human activities. But even with such disturbances all rivers continually attune their form in order to move their sediment loads downstream in an energy-efficient way. It's all just mindless physics, but the idea that a river *strives* for some higher purpose by modifying itself and its local environment is, for me, almost evocatively human and a revelation into the inner workings of the natural world.

Tireless Toil vs. Fire and Ice

Blithely unaware of Mackin's seminal work, it was remote sensing technology that reinforced these principles for me during one of my first research projects as a young physical geography professor at UCLA. For two terrifying weeks in early October 1996, a volcanic fissure erupted under Vatnajökull, the large ice cap covering most of southeast Iceland. The eruption melted through a half-kilometer of ice, blasting steam and ash through a gaping fissure that opened up on the ice surface. Nearly 4 cubic kilometers of meltwater flowed into a bowl-shaped topographic depression under the ice. The infilling water pressed upward on the overlying mass, eventually lifting it sufficiently to break the seal between ice and bedrock, causing the newly filled subglacial lake to suddenly pour out. The meltwater tunneled under 50 kilometers of glacier ice in an esker (a subglacial river) before bursting out onto Skeiðarársandur, the world's largest active glacial outwash plain, on November 5 and 6.

It was a massive *jökulhlaup,* the Icelandic word for a glacier outburst flood. The flood erupted from the glacier's snout and raced down the courses of two small, braided rivers that transit this vast, 25-mile-wide sandy plain to the sea. In fifteen hours, their combined discharge rose from a few cubic meters per second to approximately 53,000. The little rivers, accustomed to carrying trickles of water and sediment, were raging with nearly *four times* the flow of the mighty Mississippi River.

It was fortunate that the jökulhlaup struck an unpopulated area, because the destruction was immense. Ice chunks the size of small houses cracked off the glacier and rolled like toys in the flood. Some areas of the outwash plain were buried in up to 12 meters of boulders and gravel. Others were scoured down more than 20 meters. Two bridges and long stretches of Iceland's Route 1, the only highway circling the entire country, were simply washed away.

It was a big deal, and James Garvin, a prominent NASA planetary scientist, kindly invited me along on a field trip to see the damage. Through sheer luck, NASA had tested a new aircraft-based laser remote sensing technology, the Airborne Topographic Mapper, right down the center of the outwash plain just five months before the jökulhlaup struck. The technology, which NASA still uses today, aims a helically scanning laser downward from an aircraft to map surface elevations very precisely across large areas.

As we bounced around Skeiðarársandur in a ridiculously expensive four-wheel-drive rental car, we were stunned by the scale of the destruction. Huge craters pocked the landscape where stranded house-sized ice blocks were now melting away. Boulders were piled up in mounds or rolled for kilometers downstream. The highway had simply vanished, abruptly ending in midair above the deeply scoured plain.

Garvin was interested in the surreal landscape as a possible analog for the surface of Mars. I was interested in its rivers. NASA's fortuitous pre-flood airborne laser measurements offered a rare opportunity to assess how rivers shape landscapes: Which matters more—the rare, cataclysmic flood or the constant, everyday activities of flowing water? Given the scale of the carnage, I was sure it was the former. After collecting a few photos and field measurements with Garvin, I returned home and wrote a research grant proposal to bring NASA's Airborne Topographic Mapper back to Iceland to examine the rivers' damage and subsequent recovery from the jökulhlaup. My grand hypothesis was that the epic flood

In 1996, a volcanic eruption beneath Iceland's Vatnajökull ice cap triggered a catastrophic glacier outburst flood with a peak discharge roughly four times the flow of the Mississippi River. The flood tore across the adjacent plain, deeply scouring the landscape in some areas (photo) and burying it with sediment in others. Over the following years, these spectacular effects were largely erased by the far smaller flows of ordinary rivers. *(Laurence C. Smith)*

had irrevocably altered the two little rivers flowing across the outwash plain.

Five years later, after two NASA flight campaigns, lots of dirty fieldwork, and partaking of local Icelandic delicacies such as boiled sheep's head and putrefied shark flesh, my hypothesis lay in tatters. The airborne laser measurements, together with another new remote sensing technology called satellite radar interferometry, had proved that the largest jökulhlaup in modern Icelandic history had indeed demolished the river courses, but the impacts were short-lived. Where the flood scoured deep holes in their beds, the rivers refilled them. Where it piled up gravel, the rivers trundled it off, bit by bit, to the sea. Just four years after the flood, roughly half of the jökulhlaup's topographic impact on the landscape had already been erased.

What those rivers *really* cared about was getting back to business. Like ants moving sand, the rivers repaired their own bed slopes so that they could return to the job of transporting their sediment burdens in the most energy-efficient way. With centimetric accuracy, NASA's laser remote sensing technology had documented two damaged rivers striving to achieve a graded state. The effects of the cataclysm, while profound in the near term, were wiped away within a matter of years. Such is the power of an alluvial river's will.

Documentarians of Earth

On October 4, 1957, the Soviet Union successfully launched Sputnik 1, the world's first satellite, and ushered in the space age. America reacted by creating the National Aeronautics and Space Administration, a new federal agency with a mandate to place the first human on the Moon. In his book *One Giant Leap*, the journalist Charles Fishman traces how NASA's subsequent famous Apollo lunar missions spawned technological breakthroughs in satellites, computing, and telecommunications that are still advancing today.

Less well-known is that just two years after the launch of Sputnik 1, the United States launched a series of secret spy satellites as part of the Corona program, to begin snapping photos of our own planet from space. They carried high-quality film cameras and ejected their precious cargo using parachuting re-entry capsules that were snatched from midair by airplanes or scooped from the ocean by naval vessels. The capsules had a salt plug that would dissolve if not recovered promptly, sending the film to the bottom of the sea. The Corona images are now declassified and freely available online from the U.S. Geological Survey Earth Resources Observation and Science (EROS) Center near Sioux Falls, South Dakota.

A common misperception about NASA is that its only job is

space exploration. In fact, the agency's purview has always included the Earth. The astronauts of the early lunar missions photographed it using Hasselblad cameras aimed out their spacecraft windows. Unlike the secret Corona images, these photos were publicly released and generated great excitement among the general public and scientists. In the late 1960s, with enthusiastic backing from William T. Pecora, director of the U.S. Geological Survey, NASA began a historic Earth-observing program alongside its better-known lunar missions. Pecora died just days before NASA launched the first Earth Resources Technology Satellite, later renamed Landsat 1.

The core payload of Landsat 1 was a sensor called the Multispectral Scanner System (MSS). Unlike film cameras, MSS collected digital images that could be broadcast to satellite dish receiving stations. This exponentially increased the number of images that could be collected. Also, the MSS technology acquired not one but four images of the Earth's surface, each sampling a different interval of the electromagnetic spectrum (green light, red light, and two different ranges of infrared light). This enabled use of digital image processing software to combine these four independent images in various ways, producing colorful, informative digital maps of the Earth's surface.

Landsat 1 survived for six years and was a huge success. It was followed, in turn, by Landsat 2, Landsat 3, Landsat 4, Landsat 5, Landsat 7, and Landsat 8 (Landsat 6 crashed into the Indian Ocean upon launch). Landsat 9 is scheduled for launch in December 2020. With more than 7 million images compiled over five decades, NASA's Landsat program has quietly assumed the role of the longest-running documentarian of the Earth.

A good example of the power of incessant satellite monitoring was a 2016 *Nature* paper written by Jean-Francois Pekel, Andrew

Cottam, and Alan Belward from the European Commission Joint Research Center, and Noel Gorelick from Google Switzerland. They used Google Earth Engine, a cloud-based digital image processing platform, to analyze the *entire global archive* of Landsat images to track changes in the world's rivers, lakes, and wetlands continuously from 1984 to 2015. Other scientists, including me, had used Landsat satellite images to study surface water conditions for a few snapshots in time, but the 2016 study was very different. Through cloud computing, the authors ingested every picture that the Landsat satellites had acquired, around the entire planet, for thirty-two years. The sheer data volume was unprecedented.

The *Nature* paper went live during the annual meeting of the American Geophysical Union in San Francisco. With more than 25,000 attendees, it was (and is) the largest gathering of earth and space scientists in the world. Pekel was there to showcase the work, and the audience audibly gasped as his colorful global maps raced one after another across the giant projection screen. Some 90,000 square kilometers of surface water had disappeared, nearly triple the area of Lake Baikal. More than 162,000 square kilometers that had once been permanent, stable water bodies weren't stable anymore. Especially hard hit were the Middle East and Central Asia, due to extensive river diversions, water withdrawals, and drought. The satellite maps also revealed appearances of many new water bodies in places previously devoid of water. Totaling 184,000 square kilometers—roughly half the area of Germany—most were artificial, the creations of newly dammed river reservoirs. The room was buzzing. No one had ever seen so much information absorbed by a single study before.

Soon thereafter, more global surface water studies began breaking. At the University of North Carolina, George Allen and Tamlin Pavelsky merged the Landsat satellite image archive with painstaking field measurements. By hiking upstream to the uppermost headwaters of seven different rivers and measuring their channel widths along the way, the duo discovered something peculiar: The average size of a river's first-appearing headwater stream is 32 centimeters wide (plus or minus 8 centimeters), regardless of location, terrain, climate, or vegetation. It appears that all river basins extend their feeder streams upstream, like tentacles, until they shrink to the width of a dinner plate.

The physical reasons for this are still under study, but the discovery has important implications for a host of biogeochemical and sedimentary processes that occur at the knife edges of river headwater basins. In particular, open water is a significant natural source of carbon dioxide and methane greenhouse gases, so the discovery of billions of tiny feeder streams signified that the level of these gas emissions must be higher than previously thought.

These tiny headwater streams are too small and too abundant to be mapped from space. To estimate their total global surface area, Allen and Pavelsky again tapped the global Landsat archive, this time to map out the widths of every river in the world visible from space. Then, using statistics and the 32-centimeter limit revealed by their previous study, they appraised the global surface area of rivers, taking headwater feeder streams into account. They found that river water covers some 773,000 square kilometers (0.58 percent of the world's non-glaciated land surface), a number 44 percent larger than earlier estimates. Rivers therefore play a larger role in greenhouse-gas emissions than previously thought, a finding made possible by fusing detailed field studies, high-end computing, and the global archive of satellite images.

NASA's Landsat satellite program is not alone. France's series of SPOT satellites dates back to 1986. NASA's two huge Terra and Aqua satellites, bristling with sensors, date back to 1999 and 2002, respectively. Since 2014, the European Union's Copernicus program has used satellites (called Sentinels) to collect space imagery, and in situ networks (e.g., weather stations, air quality sensors, ocean buoys, irrigation water demand) to simultaneously collect measurements on the ground. This EU program has launched six satellites already, with almost twenty more planned by 2030.

Alongside these long-running government programs, very high-resolution satellite images are now being collected by private companies like Maxar (formerly DigitalGlobe) and Planet. Maxar, an international company headquartered in Colorado, owns and operates a series of commercial satellites that are mapping the Earth's surface with spatial resolutions as fine as 30 centimeters, allowing crisp detection of cars and even people. The San Francisco–based company Planet is launching hundreds of "CubeSats"—cheap, tiny satellites about the size of a breadbox, holding little more than a camera—to amass breathtaking volumes of data. The company envisions a near-future in which high-resolution satellite imagery will become accessible everywhere, every day. In 2018, the company opened a new manufacturing facility capable of producing forty new CubeSats a week. For her Ph.D. dissertation, Sarah Cooley, a student I advised, pioneered a way to use machine-learning algorithms to absorb this torrent of high-resolution satellite imagery to track surface water changes across the surface of the Earth.

These growing global archives hold petabytes—soon to be exabytes—of satellite images. Until recently, it was impractical to analyze such data volumes in their entirety. If the entire print collection of the United States Library of Congress were scanned, for example, it would comprise around 0.01 petabytes of data. With more than 8 million images spanning more than four decades, the Landsat image archive alone contains 1 petabyte. But cloud-based computing is enabling these huge, growing big-data archives to be analyzed at a rate and geographical scale quite unimaginable even a few years ago. Today, any reasonably adept person with a fast internet connection has the power to step into a time machine via satellite images, to view the past or present anyplace in the world.

The value of this big-data explosion for river studies is endless. Satellite images of flooded areas are now used to adjudicate insurance claims and improve flood risk models. Monitoring rivers upstream of population centers allows early warning of floods days before they strike. Small changes in river color are used to track suspended sediments, algae, and other colorful water-quality indicators. Thermal imaging technologies are used to map small differences in water temperature, detecting industrial effluent and locations where groundwater springs enter rivers, for example. Colin Gleason, a civil and environmental engineering professor at the University of Massachusetts, used archived satellite images to discover an intrinsic characteristic of rivers called "at-a-station hydraulic geometry," which enables reasonably good estimation of river discharge from space with no ground measurements whatsoever. These and other exciting remote sensing technologies now offer unprecedented opportunities to study and monitor the world's rivers, regardless of their remoteness or political jurisdiction.

Put on Your 3-D Glasses

Very soon, these global, rapidly growing big-data archives will consist not only of two-dimensional images but three-dimensional ones as well.

Until relatively recently, high-resolution digital topographic data were coveted and hard to find. They were produced mainly by militaries, for sensitive purposes such as navigating low-flying cruise missiles through terrain or intersecting with seismic wave recordings to detect secret nuclear bomb tests. In graduate school, I recall once missing a Thanksgiving holiday trip home because I had to tediously digitize, by hand, all of the little brown contour lines on a paper topographic map. When a fellow student secretly copied and tried to sell a restricted digital topography dataset loaned to our university by the National Geospatial-Intelligence Agency (then called the National Imagery and Mapping Agency), he was arrested and faced prison time.

Fast-forward to 2020, and vastly superior high-resolution 3-D digital topographic data are freely available online. This trend can be traced back to 2000, when NASA's Space Shuttle *Endeavour* flew a special mapping mission called the Shuttle Radar Topography Mission (SRTM). Radar (an acronym for *ra*dio *d*etecting *a*nd *r*anging) uses one or more antennas to emit and receive microwaves, a much longer wavelength range of the electromagnetic spectrum than visible and infrared light. SRTM used an amazing technology called radar interferometry, which fires two radars at once and triangulates their respective echoes to map out topographic elevations across the Earth's surface. To accomplish this, *Endeavour* flipped upside down and opened its cargo bay with one antenna aimed at Earth. A second antenna, mounted to the end of a 60-meter-long collapsible boom, was gingerly extended off the space shuttle's side. Fortunately, everything worked, and for the next ten days the space shuttle's astronauts feverishly recorded the radar data on magnetic tapes (how quaint!) as the pair of triangulating radars swept around and around our planet. By the end of the ten-day mission, enough radar data were collected to create the most comprehensive high-resolution topographic map ever made of the Earth's surface, blanketing it between 54 degrees south and 60 degrees north latitude.

Today, *Endeavour* is on display in the California Science Center museum in Los Angeles, and the collapsible boom hangs in the Smithsonian National Air and Space Museum's Steven F.

Udvar-Hazy Center in Virginia. But the SRTM radar data continue to be used and reprocessed, with many refinements and re-releases through the years. The mission's 3-D images are a standard backdrop in Google Earth mapping software and are used for everything from scientific studies to siting cell phone towers to video game graphics.

~

Radar interferometry, the technology behind the Shuttle Radar Topography Mission, can also be accomplished using a single antenna if it returns at least once to image the same area on the ground. Any passage of time between the two visits thus opens the possibility of mapping subtle *changes* in topography, in addition to the topography itself. My former graduate school roommate Doug Alsdorf, now a professor at the Ohio State University, pioneered this technique over rivers, using radar data from another space shuttle mission called SIR-C / X-SAR.

By triangulating radar echoes bounced off the flooded Amazon River, collected twenty-four hours apart from the space shuttle, Alsdorf mapped centimeter-scale drops in the flood's water level that occurred during the twenty-four-hour period between the two orbits. In response to a 12-centimeter drop in the Amazon River, he found that the surrounding floodwaters dropped 7–11 centimeters close to the river but just 2–5 centimeters farther away.

This discovery revealed that flooded river floodplains do not fill up and down uniformly, like a bathtub. Instead, flood levels (and flow directions, as it turns out) follow intricate spatial patterns caused by backwaters, vegetation, and secondary channels. Using the Japanese Earth Resources Satellite (JERS-1) and other radar satellites, follow-up studies demonstrated that river floodplains are surprisingly complex, typically piling up more water in some areas and less in others. Mapping these 3-D patterns from space provides valuable information for conserving wetlands and accurately assessing flood risks on large river floodplains.

This sort of radar technology is on the brink of going mainstream. In 2017, NASA dispatched nine different aircraft across some of the most remote areas of northern North America as part of its Arctic-Boreal Vulnerability Experiment (ABoVE), a decade-long program that is testing new sensor technologies from the air, ground, and space to answer a host of scientific questions about the changing Arctic and sub-Arctic. As the flight project leader for one of these new technologies, my job was to deploy AirSWOT, an airborne experimental radar interferometry sensor built by the NASA Jet Propulsion Laboratory, across some 28,000 kilometers of Arctic and sub-Arctic North America. The instrument package on the AirSWOT airplane included multiple radar antennas specifically designed to use radar interferometry to map water levels in rivers and lakes.

Throughout that summer AirSWOT imaged more than forty thousand lakes and rivers while I organized field teams simultaneously on the ground. Nearly three dozen people from four countries and fifteen institutions used boats, float planes, and helicopters to fan out through remote areas of Saskatchewan, the Northwest Territories, and northern Alaska underneath the AirSWOT overflights. Our main objective was to use precise field surveying equipment to test whether radar interferometry really can measure water levels as sensitively as hoped. By comparing our field surveys with the AirSWOT images collected overhead, we confirmed that this new technology can indeed map water levels across very large areas in precise and useful ways. For example, Lincoln Pitcher, one Ph.D. student I advised, used AirSWOT images to reveal the effects of subsurface permafrost on water levels across the Yukon Flats, a vast and ecologically important wetland system along the Yukon River in northern Alaska. Two others, Jessica Fayne and Ethan Kyzivat, conducted related studies using AirSWOT images collected all across western Canada and Alaska.

NASA's Arctic-Boreal Vulnerability Experiment is just the beginning. At the present time, less than 1 percent of water levels in the world's millions of freshwater lakes are monitored. River and reservoir levels are either unmeasured or are guarded state secrets in much of the world. NASA did not build AirSWOT as an end unto itself, but rather as a pilot demonstration for a new kind of satellite to be launched into space. It is a prototype for an even grander technological advance that is nearly upon us.

Big Data Meets Global Water

The new satellite, called the Surface Water and Ocean Topography (SWOT), will use radar interferometry to map changing water levels and slopes of the world's oceans, rivers, reservoirs, and lakes (see color plate). Its technology builds on a long and successful legacy of oceanographic radar satellites called altimeters, but because SWOT uses interferometry, its images will have much finer spatial resolution than traditional radar altimeters, thus enabling mapping of small inland water bodies as well as oceans. The mission is a partnership between NASA and the space agencies of France (CNES), Canada (CSA), and the United Kingdom (UKSA), and between scientists from two communities (hydrology and oceanography) and many nations. SWOT's progression from idea to reality encompasses two decades, dozens of international conferences, thousands of jobs, and an investment of more than $1 billion in public funding.

Soon after SWOT launches in 2022 it will commence mapping the Earth's surface water in 3-D at least once every twenty-one days. Over land, its measurements of water surface elevation and slope will be collected for rivers 100 meters or wider and for lakes as small as 250 by 250 meters. The satellite's preliminary measurements will be immediately posted online, followed by a reprocessed, quality-assured global dataset at least annually. All of these data will be freely available online for scientific and commercial purposes.

SWOT will also provide remotely sensed estimates of river discharge. In America, the U.S. Geological Survey maintains more than eight thousand river gauging stations and posts their data online, but such transparency is relatively rare. Outside the United States, Canada, Brazil, and Europe, river gauge data are typically sparse or secret. Water level measurements in reservoirs are even rarer. This makes monitoring river flows, and compliance with transboundary water-sharing agreements (see Chapter 2), challenging or impossible for much of the world. By providing real-time measurements online, SWOT will transform the ability of water planners, governments, NGOs, and the private sector to monitor the status of critically important freshwater resource stocks anywhere in the world.

Because SWOT is an experimental technology, it is difficult to anticipate all the ways these data will be used. At the present time, water levels in lakes and wetlands are mostly unmeasured around the world. Mission planners anticipate benefits to river commerce by supplementing sparse data from gauging stations and helping floodplain communities and businesses better protect themselves against floods. Water planners foresee tracking water storage in reservoirs and building better computer models to forecast crop yields, floods, and droughts. If even some of these aims are realized, SWOT will materially benefit humanity and the mission will be a smashing success.

By the time it launches in 2022, SWOT will have preoccupied two decades of my life. I have been involved with conceiving and planning this mission since its inception, and some of its core ambitions trace to my Ph.D. dissertation in the mid-1990s. My first-ever presentation at an American Geophysical Union conference, as a tender new doctoral student, was a talk titled "Can We Measure River Discharge from Space?" I recall feeling overwhelmed as the escalators lowered me into a milling crowd in San Francisco's cavernous Moscone Center, and then nervously presenting my ideas to a skeptical audience. After all, in 1994, the very notion of using satellites to measure river flow was absurd.

But if this exciting new technology works as planned, SWOT will be just the first of many such satellites that help us steward the world's freshwater resources moving forward. It will join the ranks of many other successful satellite missions that monitor other elements of the global water cycle—for example, the Global Precipitation Measurement (GPM) and CloudSat satellites that measure rainfall, the Soil Moisture Active Passive (SMAP) satellite that measures soil water content, and the Gravity Recovery and Climate Experiment (GRACE) and forthcoming NASA-ISRO Synthetic Aperture Mission (NISAR) satellites that detect areas of groundwater depletion (among many other things). These water-sensitive technologies supply critical observations and are also used to drive hydrological models, improving their forecasting value for everything from water-resource planning to flood risk assessment.

Together with observational data from cheap ground-based sensor networks, global data streamed from satellites will be sifted using artificial intelligence algorithms to monitor the din of the Earth's water cycle. Sensors, satellites, and models are bringing humankind ever closer to the impossible: a continuous real-time inventory of the world's water, and its changes over space and time.

~

Correctly understanding the world's water cycle is a problem that has chafed us since biblical times. As discussed in Chapter

1, Thales of Miletus and ensuing legions of natural philosophers pounded their heads over the source of the annual Nile River floods coursing through the desert. The author of Ecclesiastes 1:7 (traditionally said to be King Solomon) wrote, "All the rivers run into the sea, yet the sea is not full; unto the place from whence the rivers come, thither they return again." But how? It was an enigma.

Aristotle believed that rivers originated inside underground caves, through the transformation of air into water. In the Middle Ages, it was widely believed that river discharge to the oceans was balanced by a hidden system of tunnels that returned seawater back to the land. This idea persisted well into the Renaissance, when it was bought by none other than Leonardo da Vinci. He reasoned that just as human arteries pump blood away from the heart and veins return it, the Earth must contain subterranean veins carrying seawater to inland springs, streams, and rivers to begin the cycle anew. The many flaws in this logic (How was seawater pumped to a higher elevation on land? How did it change from salty to fresh?) seemed not to trouble da Vinci or the many others who held this view. Remarkably, the correct answer wasn't recognized until 1674, when a Frenchman named Pierre Perrault definitively proved that precipitation is the primary source of a river's flow.

Born into an educated family, Perrault was overshadowed by his two younger brothers, Claude and Charles. Claude Perrault was a successful anatomist and architect who also happened to cofound the French Academy of Sciences, translate Vitruvius' landmark *De architectura* (see Chapter 1), and design part of the Louvre. Charles Perrault found fame from penning a book of children's stories called *Histoires ou contes du temps passé, avec des moralités* ("Stories and Tales of the Past, with Morals"). In English, it is known today as the *Mother Goose* collection of fairy tales.

Pierre Perrault, after a stint as a tax collector that landed him in bankruptcy, took an interest in the Seine River. By measuring the amount of precipitation falling on land and then comparing it with discharge measurements collected in the river, he proved that rain and snow alone could account for most if not all of the flow in the Seine. His quantitative approach to the problem, favoring measurements and math over anecdotes and artwork, was part of a broader evolution in science at that time. His book *De l'origine des fontaines* ("On the Origin of Springs") put to rest a question that natural philosophers had been debating for more than two millennia—and founded the field of quantitative hydrology.

Compared to the Louvre, "Little Red Riding Hood," and "Cinderella," Pierre Perrault's legacy is largely forgotten today. Nonetheless, it set the stage for great advances in our understanding and control not only of rivers, but of the water cycle in totality. He founded a new kind of empirical, measurement-based science that sensors, satellites, and models are continuing to advance today.

Of course, quantitative hydrology has progressed greatly since Perrault's time. Scientists have identified all the major components of the water cycle, namely precipitation, subsurface flow, river discharge, evaporation, transpiration from plants, condensation, and (over snow and ice) sublimation. We know that precipitation *mostly* ends up in the ocean, but there are also large storage lockers—especially glacial ice and groundwater aquifers—that significantly prevent or delay some water from getting there. From both oceans and land, water evaporates into the air and is carried aloft for a few days before falling again as rain or snow to complete the cycle.

As water circulates through this global system it is neither created nor destroyed, but rather transformed in phase state and shuttled from one place to the next. Like a room of spinning flywheels of different sizes and speeds, some water cycles very quickly (e.g., in the atmosphere, flowing through rivers), some slowly (in groundwater, lakes, and snow), and some very slowly indeed (in deep fossil aquifers, the deep ocean, glaciers and ice sheets). On continents, it is the small but fast flywheels—water vapor, rainfall, and surface water—that rule the land and most life on it. At any given moment these compartments hold minuscule volumes of water but they circulate quickly, allowing their constant use and reuse by living organisms.

Accompanying the transformations between the phase states of water are vast releases and intakes of energy. As liquid water evaporates into gaseous water vapor, it absorbs heat from the local environment (this is how drying sweat cools skin). When the vapor rises to a cooler elevation and condenses into raindrops, the latent (stored) heat is re-released into the air, energizing storms, pressure systems, and weather. When a hurricane moves ashore it becomes cut off from its main energy source—heat released from evaporated seawater as it rises and condenses—and begins to weaken and die.

Racing through this cycle, like supercharged fuel lines, are the rivers. In terms of absolute storage capacity, they are fleetingly small, holding perhaps 2,000 cubic kilometers of water at any given instant in time. For comparison, the total volume of fresh water stored on Earth—mainly in glaciers, ice sheets, and groundwater aquifers—is around 1.4 billion cubic kilometers. But such comparisons are indeed like comparing the volume of a fuel line to the volume of a gas tank. It's the fast, concentrated throughput of water and energy that makes rivers so special, and is the prime reason that humans tend to settle next to rivers instead of lakes.

Furthermore, rivers carry *fresh* water, on a planet where almost 98 percent of all water is salty and unfit for drinking or irrigating crops. Rainfall is too diffuse to be readily harnessed. Rivers are thus tremendous physical concentrators of freshwater mass and energy, making them critical supporters of human civilization and biological life.

The Power of Models

With sensors and satellites supplying global observations of the various components of the global water cycle, it also becomes feasible to code computer models to study and predict it.

Hydrological models are powerful tools that benefit society on a daily basis. They are used to guide water utilities, schedule dam releases, and forecast floods. They inform farmers and relief agencies about droughts and help water planners adapt to short-term weather and long-term climate change.

Hydrological models are similar to weather forecasting models, which simulate the movements of water and energy through the atmosphere, but differ in that they simulate the movements of water (and energy) on land, after it falls from the sky as rain or snow. The two are often paired together, with output from weather models used as input for hydrological models.

Hydrological models range from quite simple to very complex. They use different approaches and operate over a spectrum of geographical scales. "Physically based" models try to explicitly simulate processes (like evaporation and infiltration of water into soil) from first principles. Simpler, empirically based models use real-world measurements to account for such processes, while sidestepping their physics and details. Both approaches have strengths and limitations and in practice most hydrological models are a hybrid of the two.

Like all software, hydrological models are evolving tools that are never truly finished. As scientists discover real-world phenomena (through field studies, for example), they incorporate them into the models. Often, it is the *discrepancies* between models and real-world observations that prompt new discoveries. For example, if measurements identify a decline in river discharge, but a hydrological model is unable to reproduce that decline, then something important may be missing from the model. Frustrated, its developers start testing out ideas. Perhaps reforestation—because trees transpire water into the air—could be reducing runoff? Perhaps groundwater pumping is causing headwater springs to dry up? Into the model the ideas go. Eventually a missing real-world process is discovered and coded and a new version of the software released. Field experiments are conducted to confirm that the model has indeed improved. The outcome of all this is not just a better model, but improved scientific understanding of how the natural world truly works.

In this way, modelers and field scientists have cajoled each other forward for decades. Our scientific understanding of the Earth has improved tremendously as a result. And what all models need—for development and testing, for calibration, to drive simulations, to discover missing processes, to test newly updated codes—are real-world observations.

This is why the new data firehose from autonomous sensors, drones, airplanes, and satellites is so exciting. Like their weather forecasting cousins, hydrological models need real-world observations to work well. With observations, the models grow more skilled, useful, and powerful. With observations, machine-learning algorithms can be trained. This trend has long been underway in computer science and is now gaining inroads in the Earth sciences as well. One can only wonder how Thales, da Vinci, and Perrault would marvel at today's proliferation of sensors, satellites, and models, and the coming golden age for hydrologic science.

~

The last time I saw Alberto Behar was two days before he died. He drove to my house to show me his latest idea for our coming Greenland expedition: a long-range, fixed-wing drone that would map large areas of the ice sheet, for hours, virtually unattended.

He described how it would patrol autonomously overhead while he and I were down on the ice, launching his floating ring-shaped sensors into the rushing blue rivers. As the floats uplinked their measurements to Iridium communications satellites, his drone would use GPS navigation and a camera to trace their journey, while also mapping thousands of smaller meltwater freshets gushing over the ice all around us. By flying the drone over the same area twice, from two different vantage points, we would use digital stereo-photogrammetry to create ultra-high-resolution 3-D topographic maps, to quantify the downward melting of the ice sheet surface.

I was thrilled. All of this would greatly aid scientific interpretation of the uplinked floating sensor data and help extend their findings to a broader area of the ice sheet. And for just $5,000 in components and parts, Alberto's proposed drone was a bargain at the time. Just getting my field team on and off the ice sheet would cost ten times that much in helicopter flights alone.

We shook hands on it. Visibly excited, Alberto jumped into his black Porsche and peeled away up my quiet street, heading to the airport, and home.

Within three short years, the radical ideas of this elite NASA engineer who built instruments for Mars were already humdrum. Today, all my graduate students are busily ordering components and building their own autonomous sensor packages and drones. The entry cost keeps falling, and the hardware and sensor capabilities keep rising. Space agencies and private companies are filling the skies with new technologies that the world has never seen, and giving away the data for free. Long-running satellite programs are silently filming a half-century-long movie from above. New Behars young and old are advancing the natural sciences using sensors, satellites, and models in ways unimaginable twenty years ago.

Our ongoing revolutions in technology and information are extraordinary. They will lead to vastly improved understanding of rivers and all other freshwater resources on Earth.

CHAPTER 9

RIVERS REDISCOVERED

You never can tell what goes on down below!
This pool might be bigger than you or I know....

DR. SEUSS, *McELLIGOT'S POOL*

I have a secret fishing hole and am about to do the unthinkable: share its location with you.

It is, quite literally, a hole. No one knows how deep it is, or at least I certainly don't. Its clear, tannin-stained water looks like black tea as it wells up from the ground, in a forested hollow ringed by leafy hardwood trees. Eddies boil soundlessly on its surface, proof of a vertical current rising from the depths below. Ancient walls of rock slide down into the water and disappear. A horizontal stone ledge lies perfectly positioned, like a stage, next to the boils. Barely forty feet across, the dark pool feels sacred and ageless. One imagines Mohawks once coming here to heal, or witches stealing its water for spells.

It is the chimney of a cave, through which escapes about one third of the flow of the Indian River, a smallish river in the Adirondack foothills region of northern New York State. A larger cave opening lies a couple hundred yards away; it is there that part of the river slips under the ground. It was once the entryway to a roadside tourist attraction called the Natural Bridge Caverns, a place where motorists traveling Route 3 to Lake Placid could stop and stretch their legs. They could ride a little boat into the caverns and hear about their natural history, and how they are rumored to have been a secret escape tunnel for Joseph Bonaparte, the elder brother of Napoleon and former king of Naples and Spain. Smitten by the area's natural beauty, Bonaparte once built a summer house nearby in the village of Natural Bridge, so named for the marble bedrock spanning the caverns and the flowing Indian River beneath.

My own interest in the hole was fishing. As a downtown Chicago city kid dispatched to summer vacations at Grandma's in the late 1970s and early 1980s, I had never seen anything like it. Apart from its mysterious look, "the Sinkhole," as it was known locally, was a veritable fish factory. Every morning, I would ride a borrowed bike to the factory and catch two smallmouth bass without fail. My grandmother and I ate them, so I reckon I pulled at least fifty fish out of the Sinkhole during each of my summer vacations. And I was just one of a bunch of kids who fished the place. The footpath leading down to the pool was a deep rut of packed dirt. The stone ledge was littered with empty bait cans, snarls of monofilament line, and other trash left by a more careless generation. We fished every day and the Sinkhole just kept giving. To my great envy, another boy once hauled from its depths the hugest brook trout I had ever seen—a beautiful blue-jawed, fire-bellied male stippled in moon-colored spots, over sixteen inches long—right in front of me. My competitor had threaded a heavy steel bolt onto the end of his line, to sink his bait deep into the maw of the underwater cave.

Today, the path is bushy and evidence of fishing activity has disappeared. There are few signs of humans around the glade, and hardly anyone seems to know the Sinkhole is there. I recently visited Natural Bridge and saw plenty of kids in town, but they were sitting on their porches studying tablets and phones. I was keen to return to the Sinkhole but had somehow neglected to bring a rod. Something has changed, even in me, such that I'm willing to share the location of the best fishing spot on Earth. Its coordinates are 44° 4′ 16.29″N, 75° 29′ 38.77″W.

An Unnatural Separation

I was surprised to discover that such a popular spot had fallen into disuse. I shouldn't have been—it is part of a broader pattern.

I obtained historical records of fishing license sales in New York State from the U.S. Fish and Wildlife Service, which has tracked the number of fishing and hunting licenses sold all over the country since 1958. I also obtained furbearer trapping license sales records from the State of New York. From these data, I learned that those memorable Adirondack summers of my youth in the early 1980s were actually a time of peak public interest in the outdoor recreational sports of fishing, hunting, and trapping in America.

At that time there were approximately 900,000 licensed anglers in New York State. The total number is similar today, but after accounting for population growth, the ratio of anglers has fallen from 1 in every 18 residents to 1 in 22 today. Hunting has declined in both absolute and per capita terms, with nearly 800,000 licensed hunters in the early 1980s falling to around 550,000 today. Factoring in population growth, the hunter ratio has dropped from 1 in every 22 residents when I was a boy to 1 in 35 today. Trappers, always a scarce and solitary bunch, have roughly halved in absolute numbers, falling from around 30,000 in 1980 to 15,000 today. In recent years, fewer than 1 person in 1,000 has purchased a trapping license in the state of New York.

These state-level numbers mirror a national and global trend. Some 29.8 million fishing licenses were sold in America in 2018.

This beaver trapper in northern New York State counts among an increasingly rare number of people engaged in the pursuit that once opened North America to European trade. Recorded sales of hunting, fishing, and trapping licenses, visitations to national parks, and other numeric data reveal that humans' outdoor activities peaked in the 1980s and have been sharply declining ever since. *(Laurence C. Smith)*

That is similar in absolute number to the early 1980s, but after accounting for population growth, the per capita ratio of anglers has fallen from 1 in 8 Americans to 1 in 11. The absolute number of hunters has declined slightly, from 16.3 million in 1980 to 15.6 million in 2018, with the per capita ratio crashing from 1 in 14 to 1 in 21.

Similar trends are reflected in the number of days that people camp, backpack, hike, or visit public lands in the United

States, and the number of days people spend visiting national parks in Japan and Spain. Statistical studies of these and other historical records indicate that per capita participation levels in a wide range of outdoor recreational activities peaked sometime between 1981 and 1992 and has been steadily declining ever since.

Per capita declines are easy to miss, because the populations of most countries are still growing. The absolute number of visits to U.S. national parks, for example, continues to break all-time attendance records. Their facilities and staff are straining under the load. But after factoring in population growth, we learn that the popularity of most U.S. national parks peaked back in 1987 and has been falling ever since.

The underlying reasons behind this trend are debated, and very likely still evolving. Numerous studies point to our growing preoccupation with indoor entertainment. This began with television and videos and has since progressed to internet use, social media activities, and online gaming. Regardless of why, multiple lines of independent evidence confirm that a massive global retreat of *Homo sapiens* from the natural world is under way.

These include a fascinating trend in how nature is depicted in film, language, and art. For example, one analysis of seventy years of Disney and Pixar animated feature-length films discovered a sharp departure of natural outdoor settings from the world's most beloved children's movies. When Walt Disney released *Snow White and the Seven Dwarfs* in 1937, a nature-rich backdrop was used for virtually all outdoor scenes, a convention that persisted for the next forty or so years of Disney films. But since the early 1980s, half of all Disney/Pixar animated movies have shown virtually no trace of nature in their outdoor scenes. Furthermore, even when natural settings are portrayed, they contain fewer wild animals, look less wild, and increasingly depict landscapes modified by humans.

Another fascinating study used a data-mining algorithm to track the frequency of words referencing nature (e.g., *sun, flower, rain,* and so on) in *all* of the world's English-language fiction books, film story lines, and top-100 song lyrics produced over the past century. It found a dramatic decline in the use of such words throughout popular culture. "Across three genres of cultural production," wrote the study's authors, "we have found converging evidence that the space taken by nature has been dwindling in the collective imagination and cultural conversation after the 1950s." This does not necessarily mean that people *care* less about nature, but it does raise some thought-provoking questions. "Do people conceive of nature in more utilitarian terms today than before, and less in aesthetic or spiritual terms?" the authors wonder. "Do they see nature less as something to experience and more as something to consume or control? And what do these different attitudes toward nature imply for conservation efforts and overall human well-being?"

These are all great coffee-shop questions that for now remain unanswered. But as to the underlying cause of this trend, the authors are unequivocal: The pattern is consistent with humanity's global shift toward indoor recreational activities. This trend offers numerous benefits for entertainment, education, and social interaction but does come at a cost, as we shall see next.

Nature and the Brain

Richard Louv, a journalist and children's advocate, published a bestselling book in 2005 called *Last Child in the Woods.* It assembled a growing body of psychiatric and physiological research showing that exposure to the outdoors is a critical ingredient for children's developmental health. Citing numerous scientific studies, Louv traced an absence of childhood interaction with nature to a dizzying array of problems, including attention disorders, obesity, depression, and other maladies. He coined the term "nature-deficit disorder" to collectively describe them. The book launched an international movement called Leave No Child Inside and a new organization called the Children & Nature Network (with Louv as a co-founder), both seeking to reintroduce children to outdoor activities. Its ideas are so persuasive that my wife and I included them in our recent decision to move from Los Angeles to less-urban New England after questioning why we were telling our three small children they couldn't pick up sticks and flowers in L.A.

Louv's follow-up book, *The Nature Principle,* synthesized a growing body of evidence that nature-deficit disorder extends also to adults.

A dip into this research literature is quite interesting. Three years after publication of *Last Child in the Woods,* a team at the University of Michigan documented how sending adult study subjects for a solitary fifty-minute walk through an Ann Arbor park noticeably restored their cognitive skill, whereas a walk through the city's busy downtown degraded it. The improvement in adult brain function was observed regardless of a person's mood, weather conditions, or other external factors. The researchers concluded that "Nature, which is filled with intriguing stimuli, modestly grabs attention in a bottom-up fashion, allowing top-down directed-attention abilities a chance to replenish." Sharper stimuli like car horns, stores, and traffic, capture our attention more fully, requiring directed attention to overcome or ignore. This effect appears to make urban environments less restorative to adult brain cognition than natural ones.

Importantly, peacefulness alone, such as sitting in a quiet room, could not reproduce the observed cognition benefits. Something about the *kind* of stimuli found in natural settings is required. The neurology of this likely involves the visual cortex, because study subjects won some of the same benefits simply by viewing pictures of natural outdoor settings. Whatever the reason, there is clearly something special about the modestly interesting stimuli of nature—even those found in a city park—that restores the cognitive abilities of our brains.

These benefits extend also to individual feelings of mood and self-esteem. A pair of meta-analysis studies (which apply statistical methods to the datasets of multiple other studies) led by Jo Barton at the University of Essex found strong mental health benefits of "green exercise"—activity in the presence of nature. Her research analyzed ten United Kingdom studies involving more than twelve hundred individuals and found that all green environments improve both self-esteem and mood—and that the presence of *water* increases these benefits even more.

In a book called *Blue Mind,* the marine scientist Wallace Nichols further develops this idea, arguing that human neurology responds positively not only to nature per se, but to water in particular. The book meanders through water's numerous physiological and societal benefits, focusing especially on cognitive, emotional, and psychological ones. The mere sound of running water, for example, measurably reduces stress. Noise machines mimicking rushing brooks and crashing waves help people sleep. Cancer patients who are shown videos of gurgling creeks and waterfalls experience significant reductions in the stress hormones epinephrine and cortisol, and so on. In 2011, Wallace began convening regular conferences bringing together neuroscientists, psychologists, artists, and hydrologists to explore the idea that, deep down, the human psyche is fundamentally attracted to bodies of water.

I'm no neurologist, so will accept the syntheses assembled by Louv, Barton, and Wallace at face value. From personal experience, after forty-odd years of spending time on rivers, I'll admit that their claims resonate with me. Both as a child and as an adult, I have regularly experienced a sensation of calm, clear thought that settles over my mind within thirty minutes or so of poking around a river.

It wasn't the fish that enticed me daily to the Sinkhole and my handful of other favorite fishing spots along the Indian River. It was the opportunity to pass a sustained period of time in a place that was at once predictably calm yet moderately stimulating. My fishing rod would lie forgotten as I passed an unhurried hour overturning river rocks looking for crayfish and hellgrammites. I watched eddies swirl and damselflies hunt and observed how sand pulses down a river. Those hours gave me more than just passing some pleasant summer days outside the city. They taught me focus, contentment while alone, and how to see beauty in simple things. And whether you are a child or an adult, I do believe that any river, pond, or patch of parkland can offer this.

Even in the city.

Three Moments in Manhattan

My wife was recently invited to a wedding in Manhattan, so we decided to station the kids with their grandparents and make a long weekend of it. During the course of those three lovely days, three experiences solidified for me some trends regarding urban rivers that I had been mulling for some time.

The first happened as we found our way to the wedding venue. Our subway stop was a gleaming new metro station (34th Street–Hudson Yards) in Chelsea, adjacent to the Hudson River. Most New York subway stations are cramped and grimy, but this one was airy, modern, and bright. When we emerged above ground, we found ourselves in the midst of a vast construction site, with glass high-rises around a gigantic, beehive-shaped art structure made of stairwells and landings (I later learned it was called The Vessel and conceived by the U.K.-based designer Thomas

Heatherwick). As we navigated through temporary plywood walls and the crashing sounds of heavy equipment, I was struck by how the new buildings were strategically designed to maximize sightlines of the Hudson River, which flows along the western boundary of the construction site.

The site was enormous, bounded by the river on the west, 10th Avenue on the east, and 34th and 30th Streets on the north and south. The glassy new buildings were being raised *over* Manhattan's 27-acre Hudson rail yard facility, located next to the river. The rail yard was not being demolished, but rather covered over and turned subterranean. Rising above it, on pilings placed between the train tracks, was the new skeleton of Hudson Yards (see color plate), the largest development Manhattan has seen since the 22-acre Rockefeller Center was created in the 1930s.

According to its developers, Hudson Yards is the costliest redevelopment project in U.S. history, with an anticipated total price tag of $25 billion. Like most large twenty-first-century urban development projects, it emphasizes mixed-use zoning, meaning a blend of high-density residential, commercial, and outdoor space. A new complex of apartment towers, shops, restaurants, offices, and outdoor common areas will be built, as well as a new performing arts center, hotel, and public school. Along with its own 14 acres of outdoor greenspace, Hudson Yards will connect to the High Line (an abandoned elevated track that was recently converted into an extraordinarily popular vegetated public walkway) and Hudson River Park, a newish greenspace hugging the Hudson River along four miles of lower Manhattan. The project will create 18 million square feet of floor space, some four thousand residences, and more than one hundred stores. When completed in 2022, Hudson Yards is expected to generate approximately 55,000 jobs and $19 billion annually in a newly created riverfront neighborhood. It is, without question, the most overwhelming urban redevelopment project I have ever seen.

My second experience was the wedding itself. The ceremony and reception were held inside Pier 61, a repurposed former shipping terminal at the historic Chelsea Piers.

Pier 61 juts out into the Hudson River estuary. Its enclosed, elongate space is ideal for special events, with water views and abundant natural light pouring in from all sides. The Lighthouse, as the event space is called, is located at the very end. A flower-draped *huppah* was positioned in a corner where two floor-to-ceiling glass walls converged, allowing the marrying couple and their guests to face the water. Like a storybook painting, the couple was thus framed inside a spectacular panoramic view of the Hudson, the Statue of Liberty, and the waterfronts of New Jersey and lower Manhattan.

The ceremony took place at sunset. The sky blazed orange and crimson, and the couple was silhouetted by the dying light. As if on cue, the waterfronts of New Jersey and New York began to glitter with lights. The Hudson blackened into an abyss separating the two sparkling shorelines. Rabbis intoned the laws of Moses and murmured passages in Aramaic. Between the sunset, ancient ritual, and the timeless current sliding by, it was impossible to not reflect upon the cycle of life embodied in the young couple and the people of all ages assembled around them. The ceremony would've been great even in a basement, but the physical setting—jutting out over the black water surrounded by city lights—made the moment magnificent (see color plate).

The third experience happened the next day. A longtime friend invited us to a tour and cocktails at Kings County Distillery in Brooklyn. Despite having been founded only in 2010, it is somehow New York City's oldest distillery and the first to open since Prohibition ended in 1933. Its specialty is corn-based moonshine and bourbon, distilled on site using locally sourced ingredients. The spirits are funneled into charred oak barrels and aged on the plank floor of a restored two-story red-brick building. The charred barrels and aging process darken and flavor the clear corn-based liquor into quite excellent amber-colored whiskies. The tour was fascinating and we hung around afterward for libations at the Gatehouses, the distillery's public tasting room and bar.

Like the distillery's production areas, the bar was genuine retro. It was sensitively fitted into the original brick guardhouse that once protected the entryway to the Brooklyn Navy Yard, one of America's longest-running military shipbuilding compounds. It envelops a large embayment of the estuarine East River in New York City.

As a shipyard, the Brooklyn Navy Yard operated continuously from 1801 to 1966, making it the longest continuously operating manufacturing facility in New York State. It produced many famous warships, including the USS *Monitor* (the ironclad ship that fought the CSS *Virginia* to a draw in the Battle of Hampton Roads; see Chapter 3), the USS *Arizona* (the first ship sunk in the Japanese attack on Pearl Harbor), and the USS *Missouri* (upon whose deck Japan formally surrendered to end World War II; see Chapter 4).

After 165 years of operation, the Brooklyn Navy Yard was decommissioned in 1966. Tens of thousands of workers were laid off, and the rusting, decaying navy yard became a national emblem of the loss of industrial manufacturing jobs in America. The 300-acre property was sold to the City of New York, which then reopened it as a commercial park. For years, the property's main tenants continued to be shipbuilders, but by the late 1980s they too closed down, and the dilapidated facility was shuttered.

The city then formed a development corporation to reimagine uses for the property. Over the next decade, its many large warehouses and buildings were subdivided and remodeled to attract a diverse assortment of small businesses. By the early 2000s, the number of tenants leasing space in the Brooklyn Navy Yard had grown to 275, including a rapidly growing number of green manufacturers. Steiner Studios, a film and television production campus, opened in 2004 and expanded again in 2010 and 2017. Kings County Distillery relocated their whiskey operation inside the old Paymaster Building, the aforementioned historic brick structure that had once been the navy yard's bank. By 2018, there was a 250,000-square-foot Green Manufacturing Center, a 35,000-square-foot greenhouse, and a 65,000-square-foot rooftop farm. The Brooklyn Navy Yard had been transformed from a boneyard of lost manufacturing jobs to a thriving hub of diverse, forward-looking urban businesses.

~

These three seemingly unrelated stories illustrate some broad economic and environmental trends that are transforming urban riverfronts in cities around the developed world. Recall from earlier chapters the ancient relationship between cities and rivers, which dates to our earliest agricultural civilizations. The nature of that relationship has changed time and time again over the millennia. Now, the way that cities connect to rivers is changing once more. In fits and starts, our cities are moving away from a pragmatic, industrial relationship with rivers to one in which we increasingly view them as sources of human well-being.

A decades-long migration of heavy industry, shipbuilding, and manufacturing industries to developing countries resulted in the shuttering and decay of many waterfront industrial properties. This disappearance of industrial activity, together with pollution-control laws (see Chapter 6), is making urban riverfronts tolerably clean, aesthetically attractive places to live and work. Because nearly all major cities were originally founded along rivers (see Chapter 2), and many urban cores have revitalized, derelict riverfront properties in affluent cities have become ripe targets for redevelopment.

Property owners and city planners have struggled for decades to repurpose these former industrial sites, with mixed results. But today, a new generation of urban designers and planners is eyeing them in a new light. As pointed out to me rather wistfully by Liz Pulver, an award-winning New York City landscape architect who fits minuscule green oases into scarce slivers of urban land, industrial riverfront properties are typically huge compared to other available parcels. Their large size attracts ambitious redevelopment projects incorporating mixed-used residential and commercial space, affordable housing programs, environmentally sustainable materials, and lots of public outdoor greenspace.

Modern urban planning is all about public commons and greenspace, and large riverfront properties offer unusually good opportunities to create them. This is due both to long-standing legal norms preserving public access to rivers (dating to Roman times, as described in Chapter 1) and physical threats from the rivers themselves. By definition, floodplains are naturally prone to flooding and erosion, making them risky locations for permanent buildings. Such land is well-suited for use as outdoor recreational space. Furthermore, most municipalities enforce strict setback requirements along rivers, creating obvious and legally available opportunities for public walkways and easements along their banks.

In coastal areas, rising sea levels add another long-term existential threat to waterfront structures, heightening scrutiny by municipal planners, who are increasingly demanding even wider setbacks for new development. While all of this increases the cost and headaches of building near waterfronts, the wide buffers also create exciting new greenspace opportunities along the water's edge for large parcels of formerly industrial land.

Developers are rising to these challenges. In the greater New York City area alone, dozens of major riverfront redevelopment projects are being planned or built at this time. Virtually all include public parklands in a crowded city that desperately needs them. Because most seek to develop former industrial sites that have been fenced off for decades, proposals to repurpose them can realistically win support from local residents and politicians, especially when a new public waterfront park is on offer. Done right, such projects create new opportunities to connect urban neighborhoods to rivers.

Take, for example, the site of a former Domino sugar refinery, located on the bank of the East River in the Brooklyn neighborhood of Williamsburg. The plant shuttered in 2004 after a long and storied 132-year history in American sugar production. The site is now called the Domino Sugar Factory project, an 11-acre mixture of residences and greenspace. The new neighborhood will have 2,800 apartment units, with 700 reserved for affordable housing, and a 6-acre public park along the waterfront. Some historical elements of the old sugar factory will be preserved, such as warehouse columns, elevated crane tracks, and syrup tanks. Domino Park will thus create a public space for all Williamsburg residents to access grass, sports fields, and waterfront along a previously inaccessible stretch of the East River.

Two miles to the north, on the same bank of the East River, an even larger redevelopment project is under way in Brooklyn's Greenpoint neighborhood. Named Greenpoint Landing, the project will build ten high-rise towers and buildings on a 22-acre former light industrial site near the junction of the East River and Newtown Creek. It will have 5,500 residential apartments, including 1,400 affordable housing units. There will be a new

Numerous waterfront redevelopment projects are under way in Brooklyn, New York, including this major project at Greenpoint Landing, a former light industrial site on the East River. *(Laurence C. Smith)*

public school and 4 acres of public parkland installed along the riverfront.

I first visited the Greenpoint Landing construction site in 2018 and was shown around by Karen Tamir, the project's lead landscape architect, and Jovana Rizzo, a publicist. Tamir is a senior architect and designer at James Corner Field Operations, a landscape architecture and urban design firm that specializes in repurposing old industrial sites into urban parks. Field Operations (as the company is known) designed Manhattan's enormously popular High Line elevated walkway, which opened in 2009. Some of their other recent waterfront projects include Navy Pier in Chicago, Presidio Parklands in San Francisco, Central Waterfront in Seattle, Knight Plaza in Miami, Race Street Pier in Philadelphia, South Park Plaza at Queen Elizabeth Olympic Park in London, and Qianhai Water City in Shenzhen. On the East River alone, the firm's projects include Greenpoint Landing, Domino Park, and Cornell University's new technology campus, being built on Roosevelt Island right out in the middle of the river.

Like most New Yorkers, Tamir was all business, rattling off grade elevations and whipping through a sheaf of architectural drawings too fast for me to keep up. I had to beg her to slow down so I could scribble some notes. "Sandy made a big change in how we use waterfront," Tamir told me flatly. "All of our projects are now pushed in a more resilient way." Later, as we stood around in our hardhats waiting to be escorted inside a partially built tower, she glared at the surrounding diggings and rock piles and muttered that the terracing was behind schedule. I suspected someone would be hearing from her about it the second I left.

When Hurricane Sandy struck the Eastern seaboard in 2012, much of the property was flooded. Its developers, George and Marian Klein of Park Tower Group Ltd., then hired Field Operations to design a master plan for a greenspace that could withstand future storm surges and sea level rise. At that time, most of the property was just 5 to 5.5 feet above sea level. But by adding earthen fill in a series of stepped terraces, its ground elevations were raised an additional 3 to 5 feet. The new terraces will also help protect the surrounding inland neighborhood from flooding. Like the Domino Sugar Factory project, the new buildings are set back from the water, with ground floors at least sixteen feet above sea level.

Owing to the property's large size, its artificially raised elevation is quite unnoticeable. The terraces I saw being built along the riverbank will create a long, stepped public park, extending nearly half a mile along the waterfront. Like Domino Park, it will preserve historical elements of the property's industrial past, including a giant steel buoy lying half-buried in the sand. The jagged wood pilings of the site's original pier will be left in place, with a new public pier built alongside it, extending out into the East River.

Greenpoint Landing has plenty of adversaries. Many local residents oppose a ten-building megaproject towering over their quiet neighborhood of low brick buildings. While gaining a new waterfront park is nice, they dread the increased noise and traffic that will accompany the arrival of five thousand new apartments. They point to the project's proximity to sea level rise and to Newtown Creek, the toxic East River tributary draining a Superfund site swum by the environmental activist Christopher Swain (see Chapter 6).

Global Urban Renewals

Like so many other neighborhood groups that have organized to battle huge development projects, the opponents of Greenpoint Landing have been unable to halt it. Indeed, the revised zoning laws of New York City now encourage dense riverfront development. The origins of this new zoning policy go back to 1980, when 90 acres of dilapidated piers and warehouses in lower Manhattan began transforming into what is today Battery Park City. The overwhelming success of that project prompted the city to release a comprehensive waterfront plan in 1992 proposing to rezone New York's industrial waterfronts, opening them up for public access, recreation, and residential redevelopment. A first wave of zoning changes was passed the following year. In 2005 Brooklyn's Greenpoint and Williamsburg waterfronts were similarly rezoned.

In 2011, the City of New York released *Vision 2020,* an updated waterfront plan that identified even more zoning changes and numerous specific waterfront redevelopment projects throughout the city. In the document's written foreword, the city's mayor, Michael Bloomberg, praised the plan's opening of miles of waterfront that had been closed to the public for decades. He pledged to ensure "that each neighborhood has access to the recreational space that is so vital to our residents' quality of life." Together with a companion implementation document called the *Waterfront Action Agenda,* the plan prioritized more than 130 new waterfront redevelopment projects throughout all five boroughs of New York City.

One of these is Brooklyn Bridge Park. An open plaza beneath the Brooklyn Bridge connects an existing riverfront park to the north with newly repurposed piers and a larger waterfront park to the south. A million-square-foot warehouse has been converted into apartments and stores, as have seven warehouses (built as coffee storage in the late 1800s) that were largely abandoned in the 1960s. There are restaurants, a hotel, and event spaces. It is a spectacular and vibrant waterfront location offering plenty of outdoor space and views of the Brooklyn Bridge, the East River estuary, and Manhattan.

Looking north, to the confluence of the Harlem and East Rivers in the Bronx, the Mott Haven–Port Morris Waterfront Plan will repurpose the Harlem River Yards, a 96-acre former rail yard that has been closed to the public for many years. The land is publicly owned but leased to a property holding company, which in turn subleases it to a waste transfer station, a power plant, a package delivery hub, and a newspaper printing and distribution center. After rezoning, the new waterfront plan will create publicly accessible riverfront parks integrated with some thirteen hundred residential units, many of them converted from old factory and warehouse buildings. The new public outdoor spaces will include three different waterfront parks, a waterfront path connecting them, a boat launch, and a fishing pier.

New York's grandest waterfront plan, which the state's governor, Andrew Cuomo, singled out for praise in his 2018 "State of the State" speech, reimagines Brooklyn's Red Hook area, near

Governors Island. It would redevelop more than 130 acres of waterfront, on land currently occupied by a shipping container terminal, a cruise ship terminal, and the New York Police Department tow pound. The plan envisages 45,000 apartments and a variety of commercial spaces and outdoor parks. The new waterfront neighborhood would connect to lower Manhattan via a new subway station tunneling under the East River. If the Red Hook project moves forward, its scale will be immense—some six times larger than the $25 billion Hudson Yards project that so impressed me on the way to that Manhattan wedding.

~

This waterfront revolution now under way in New York City is part of a broader, global phenomenon. Other efforts to transform urban rivers from dilapidated industrial corridors to attractive public and residential spaces are now under way in major cities all around the world. Consider just five examples: London, Shanghai, Hamburg, Cairo, and Los Angeles.

In the nineteenth and early twentieth centuries, London's prominence as a center of mercantilist trade required many piers, dockyards, and warehouses to be built in the London Dockyards, along the River Thames in the city's East End. Other shipping infrastructure was built along the River Clyde in Glasgow, the Mersey in Liverpool, and the Tyne in Newcastle. These labor-intensive dockyards survived numerous recessions over the years, but the rise of standardized container shipping in the 1970s and '80s spelled their doom, and most became derelict and abandoned.

London helped write the playbook on repurposing such properties. The city founded a redevelopment corporation for the London Dockyards, an idea imitated many times since (including at Brooklyn Navy Yard). A milestone of the dockyards' long revitalization was the completion of the iconic

Canary Wharf Tower in 1991 which, together with the nearby Millennium Dome built in 1999, anchored the East End skyline as new redevelopment projects spread throughout this formerly industrial area. Other new projects are now in progress east of Canary Wharf, including the new Royal Wharf riverfront neighborhood next to London City Airport. This 40-acre redevelopment on the Thames is converting a former industrial complex into restaurants, shops, a school, outdoor parks, and more than three thousand homes and apartments.

Upstream, in central and western London, some huge riverfront projects are also under way along the Thames. A massive $17 billion megadevelopment in the Nine Elms neighborhood will be anchored by Battersea Power Station, a former coal-fired power plant, and London's largest brick building, that was fully decommissioned in 1983. Thousands of new residential units, retail spaces, and parks have already been built on this approximately 42-acre waterfront site, with plans for thousands more. There will be hotels, office buildings, two subway stations, and—of course—public waterfronts.

The former power station itself (which incidentally appears in Monty Python's *The Meaning of Life* and Christopher Nolan's *The Dark Knight* and on the album cover of Pink Floyd's *Animals*) will be totally repurposed, with much of the famous structure's brickwork and exoskeleton preserved. Apple plans to relocate its U.K. headquarters in the station's former boiler rooms in 2021, and completion of the entire Nine Elms complex is slated for 2025.

Importantly, London's new riverfront residents will soon have a cleaner Thames to enjoy. The city has a 150-year-old combined sewer system, meaning that storm runoff is routed into the same drain tunnels as raw sewage. When rainstorms flood the system beyond capacity, the excess stormflow-sewage mixture is dumped into the Thames. While dilute, this contaminates the river, making it unhealthful for aquatic life and people. This longstanding problem will be remedied by the Thames Tideway Tunnel, a 15.5-mile-long pipeline being burrowed under the river. It will use specially designed collection shafts to capture the contaminated liquid.

An interesting side benefit is that the collection shafts will be capped with platforms, creating a series of small public parks jutting out like miniature peninsulas into the Thames along some of the most prominent areas of London. The purpose of these parks, according to their lead architect, is to encourage people to visit the river. Some are even designed to allow Londoners to put their weary urban feet into the newly cleaned-up waters. With an estimated cost of more than $5 billion and scheduled completion in 2024, the project is one of the largest water infrastructure projects in the United Kingdom.

~

I first visited Shanghai in 2017. Kang Yang, a friend and professor at Nanjing University, offered to show me around the city. When I asked his opinion of the most imperative place for visitors to see his answer was immediate: the Bund.

The Bund is an esplanade along the historic west bank of the Huangpu River in downtown Shanghai (see color plate). Across the river soar the gleaming Oriental Pearl Tower, Jin Mao Tower, and recently completed Shanghai Tower, currently the tallest building in China. During the Yangtze River gunboat era, Shanghai was an important treaty port in Asia (see Chapter 3), and the Bund was its international financial center. Foreign capital lined the shore with grand, Beaux Arts–style customs houses and banks, many of them still preserved today.

By the 1990s, the waterfront had fallen into neglect and was choked off from the city by a ten-lane highway. That changed in the 2000s, when Shanghai reinvented the Bund in preparation for the 2010 World Expo. Six lanes of traffic were relocated underground, and a wide, elevated pedestrian esplanade was made along the Huangpu River. New pedestrian walkways safely connected it to the rest of the city. Outdoor plazas and pavilions were built, and new greenspaces planted with gardens and trees.

As a place for people to congregate, the project was a huge success. When I visited, the esplanade was thronged by people walking, talking, or just gazing out at the river.

Shanghai's revitalization of the Bund became part of a broader plan to rezone and redevelop both banks of the Huangpu River throughout the city. At least 45 kilometers of walkable waterfront have been opened. The Hongkou waterfront (also called the North Bund) will have a riverfront park, and the Xuhui waterfront (West Bund) has art galleries, museums, a theater, concert hall, and still more parks. The city's Pudong district (on the east side of the river) will also have numerous waterfront parks. Once a major commercial shipping hub, Shanghai's old piers, dockyards, and warehouses are now giving way to residential towers, shops, restaurants, and outdoor greenspace. Like New York, Shanghai's city planners are repurposing these former industrial sites to create a new kind of urban lifestyle along its river.

In Germany, an entirely new district is going up on the Elbe River next to downtown Hamburg, the country's second-largest city and one of Europe's busiest ports. Named HafenCity, this 388-acre former port and warehouse complex is the largest urban redevelopment project currently under way in Europe. To enable the project, the port ceded HafenCity's land and waterways. To protect against flooding and sea level rise, it is being built atop artificial mounds of *Warften,* or compacted fill, effectively raising the ground at least 8 meters above sea level.

HafenCity's first buildings were completed in 2009, its first subway in 2012, and a new university in 2014. Elbphilharmonie, a spectacular new concert hall named after the river, opened in 2017. Designed by a Swiss architecture firm, it cost nearly $900 million to build and has transformed the cultural and architectural landscape of Hamburg. When completed around 2030, HafenCity will expand the city's existing downtown area by about 40 percent, with more than 2.3 million square meters of floor space, stores, restaurants, and more than seven thousand residences. With 14 kilometers of waterfront promenades and parks along the Elbe River embankment and former docksides, virtually all of HafenCity will have views of or access to the water.

Few vacant riverfront properties exist in downtown Cairo, one of the oldest yet fastest-growing cities in the world. Much of the city's growth is in its outskirts, creating an affluent suburbia and low- and middle-income core. But a sweeping plan to revitalize the city center between downtown Cairo and the Nile River has begun. The plan was initially backed by Hosni Mubarak, Egypt's president at the time, and has since been resumed by the current president, Abdel Fattah al-Sisi.

Anchoring this redevelopment is a narrow, triangular sliver of land on the bank of the Nile River that has been chosen as the site for a soaring seventy-story skyscraper of twisting glass called the Nile Tower. Designed by the late award-winning architect Zaha Hadid, it will preside over a new 86-acre redevelopment zone in the Maspero Triangle, a recently demolished low-income neighborhood next to the Nile. Its residents bitterly fought the project for years but were ultimately evicted and their homes razed in 2018. Like the other redevelopment projects described in this chapter, it will create mixed-use residential and commercial space and an appealing waterfront esplanade lined with palm trees. There will be high-rise buildings, stores, and a walking bridge to the Zamalek neighborhood, a quiet island in the Nile River.

Unlike the other projects I have described (which emphasize

affordable housing, access to mass transit, and preservation of historical artifacts), a clear objective of this one is to create a bold symbol of modernity and affluence. The Nile Tower alone will cost upward of $600 million, and its upscale apartments will cost far more than other properties in the surrounding area. The building's lower floors are slated for upmarket amenities including a spa, night club, exclusive shops, and a casino. The project's residential side will be financed by private funding; at the time of writing, Egypt was aggressively courting foreign investors for the project.

This project appears to be more than just another real estate deal. In addition to attracting foreign investment, it aspires to symbolize a new period of Egyptian affluence and power in the region. Cairo is a center of gravity for the Middle East, and the Nile Tower and Maspero Triangle projects represent a new kind of riverfront use in the heart one of the world's most ancient and glorious civilizations, as described in Chapter 1. Should the project advance, once again a splendid structure—one of the tallest in Africa—will rise from the banks of the Nile.

Perhaps no other riverfront revitalization plan has captured a sense of rediscovery more than that surrounding the Los Angeles River.

As a resident of L.A. for over twenty years, I can attest that most Angelenos do not even know where the Los Angeles River is. Unlike most cities, L.A. did not grow along its river's banks. Unnavigable and dry for much of the year, the river was never a shipping artery or even a reliable water supply. It was prone to raging floods and occasional avulsions, sometimes jumping its banks to attempt a new course.

After a series of devastating floods in 1914, 1934, and 1938 claimed more than one hundred lives and over $1 billion in damages, the U.S. Army Corps of Engineers tamed the Los Angeles River. Its banks were carved back, and the widened channel lined with great sloping concrete abutments. For much of its 51-mile length, the riverbed itself was also paved.

Ecologically, the river was transformed from a lush belt of vegetated islands and ephemeral pools to a trapezoidal concrete channel. Societally, the river became a giant open storm sewer and was largely forgotten except by anonymous street art muralists, illegal street racers, and set location managers for a host of films, including *Grease* and *The Dark Knight Rises.* During storms, swift-water rescues are common in its raging channel, where discharges can exceed 100,000 cubic feet per second.

A grand scheme is now under way to restore some ecology to the concrete eyesore, and to reincarnate the L.A. River as a corridor of public greenspace and economic revitalization. The idea dates to at least 2002, when the City of Los Angeles drafted an early conceptual plan for the river's rehabilitation. This initial document evolved iteratively over the years, and in the process, forged a strong partnership between the city, Los Angeles County, and the U.S. Army Corps of Engineers. The globally renowned architect Frank Gehry recently signed on to help design the final blueprint for the river's restoration and development. Working closely with Gehry is River LA, a nonprofit organization whose mission is "to integrate design and infrastructure to bring people, water and nature together" along the river.

The details of a final master plan for the entire 51-mile river length have not yet been released but are expected in 2020 or 2021. In 2016 the U.S. Congress approved a major first phase, called the Los Angeles River Ecosystem Restoration project, to restore a natural riparian ecosystem along an 11-mile stretch through downtown Los Angeles. In 2017, another milestone was passed when the city purchased an abandoned freight-switching railyard called Taylor Yard River Parcel G2. This 42-acre river-front property, formerly owned by Union Pacific Railroad, is a critical part of the overall river revitalization plan. "This vast site can transform how Angelenos connect with the natural world," said the mayor of Los Angeles, Eric Garcetti, in his announcement of the land purchase, "because it will allow for habitat restoration, and open more than a mile of direct access to the river for local communities that have been cut off from it for too long."

Smelling a broader economic revitalization opportunity, real estate developers are now initiating numerous projects along this neglected waterway. At the time of writing more than twenty new development proposals were in various stages of progress along the banks of the Los Angeles River. These include projects around the so-called Bowtie parcel (another chunk of abandoned rail yard) that is slated to become an outdoor public art space, a futuristic bridge, and a riverfront park. In downtown Los Angeles, another new bridge will connect the Los Angeles River Greenway Trail (a walking and biking path) with the Glendale Narrows Riverwalk and bikeway system on the opposite side of the river. Farther downstream, a plan called the Lower LA River Revitalization Plan has proposed an astonishing 146 river-front projects, including parks, vegetated walkways, and a nature overlook running from Vernon to Long Beach.

This lower stretch of the Los Angeles River passes through South L.A.'s working-class communities of Lynwood, Compton, and North Long Beach, some of the grittiest parts of the city. The river's cracked concrete floor is barren and lifeless. The surrounding housing tracts are sunbaked and dusty, with no green-space for miles. If even a fraction of these proposed projects come into being, Richard Louv, Jo Barton, Wallace Nichols, and all supporters of the No Child Left Inside movement would surely and resoundingly applaud.

At present, the Los Angeles River is little more than a concrete storm runoff drain. A massive river restoration and revitalization plan now seeks to transform the neglected waterway into a vibrant recreational, ecological, and development corridor. *(Laurence C. Smith)*

The Urban Majority

The time is right for the trend in urban river renewals that I have described.

Our species surpassed a historic threshold sometime in the year 2008. The exact moment will never be known, but somewhere on Earth, a baby was born, and we passed into an unfamiliar world. We became urban in the majority.

Never before in the history of human civilization have more people lived in cities than rural areas. Never before have people been so unacquainted with farming or hunting or raising animals for food. Never before have so few children played outdoors. In 2008 we officially became an urban species, overlords of a global food-production economy of our own creation.

Since that baby was born, our population has grown by another billion people, from 6.7 billion to 7.7 billion. Our urban majority has grown from 50 to 55 percent. By 2050, it will become nearly 70 percent, with 2.5 billion more city slickers than the 4.2 billion alive today. The pace of this urban growth is equivalent to adding three more Shanghais to the planet every single year for the next thirty-two years.

Most of this urban growth is happening in Asia and Africa, with more than one-third in India, China, and Nigeria alone. Utterly without precedent is the emergence of many hundreds of cities with populations of 1 million people or more, and dozens of "megacities" having populations of 10 million people or more. Demographic models project that the number of megacities will nearly double, rising from twenty-eight in 2015 to nearly fifty by 2035. Should their projections hold, the world's biggest city will be Delhi, India, with 43 million people. Tokyo, Shanghai, Dhaka, and Cairo will round out the top five.

To put these changes into perspective, greater New York was the world's second-largest urban area in 1970, when there were just 144 million-plus cities globally. By 2035, it is projected to rank thirteenth, and the number of million-plus cities will rise to somewhere around 759.

What will these cities be like? The future is of course malleable, but unless something dramatic happens to upend this decades-long trend, we can expect that most of today's large cities will become even larger. We can expect even more high-rise buildings will further increase the housing density of scarce urban land, clustered around mass transit or personal mobility corridors. We can expect that societies will have a larger percentage of elderly people. We can expect skies filled with autonomous flying equipment and streets crammed with novel autonomous vehicles. We can expect urban land prices to be sky-high, and access to open green spaces and nature to be at an absolute premium.

The urban river trend I have described, of cities transforming obsolete industrial waterfronts into new corridors of public greenspace, can help with this last issue. In an urban world, few children go camping in the woods. In an elderly world, there will be many whose rock-climbing days are behind them. Brain cognition studies and common sense tell us that nature is good for people. Science and common sense tell us that removing industrial waste from rivers is good for nature. And while it is impossible to defend all floodplains and coastal deltas from the changing flood probabilities and rising sea levels wreaked by climate change, the compact size and comparative wealth of cities make them better positioned to engineer defenses against them.

After Hurricane Katrina demolished the Gulf Coast in 2005, greater New Orleans built extensive new flood defenses and the city's population rebounded. But elsewhere along the Gulf Coast, miles of erased beachfront communities have never recovered. Cities, with their large populations, high property values, and relatively small size, stand a better chance of protecting their waterfronts, at least for another century or two.

~

To illustrate the scale of the urban river opportunity, let us turn to my global GIS study with Sarah Popelka, first introduced in Chapter 2, to quantify how common these natural features are in today's cities. As you may recall, this new dataset uses a global map of large rivers (30 meters or wider, as mapped from satellite remote sensing) to quantify their use as political borders. Now, let us examine how this same global river database aligns with the spatial distribution of human populations worldwide (see table on the next page).

From this exercise, we learn that not only are we humans an urban species, we are also a river species. Indeed, nearly two thirds (63 percent) of the total world population lives within 20 kilometers of a large river. Some 84 percent of the world's large cities (defined as having a population of at least 1 but less than

10 million) are located along a large river. For the world's megacities (population greater than 10 million), the number rises to 93 percent. Because my analysis does not include smaller waterways, which are far more abundant (for inspiration see the global map at the front of this book, presenting the topographic drainage pattern of the Earth's surface), we can safely conclude that these percentages are conservative estimates.

	Total	Near Large Rivers	(%)	Near Coasts	(%)	Near Both	(%)	Near Coasts only	(%)
Cities w/ Pop. >10M	30	28	(93)	21	(70)	19	(63)	2	(7)
Cities w/ Pop. >1M	429	359	(84)	181	(42)	138	(32)	43	(10)
All Cities	75,445	42,946	(57)	9,073	(12)	3,773	(5)	5,300	(7)
2015 World Pop.	7,349,286,991	4,623,518,316	(63)	1,397,438,116	(19)	1,038,787,479	(14)	358,650,637	(5)

From this research we also see that the vast majority of "coastal" cities are actually river cities, meaning they are located on river deltas. It is an oft-repeated fact that many people live near coasts, and indeed nearly one in five (19 percent) of us do. However, most are actually living on river deltas. Just 5 percent of the world population and less than 10 percent of cities are located near coasts but not rivers. Put simply, rivers are the preferred natural features around which we have built our urban civilization. Today, nearly all of the world's major cities have the opportunity to interact with rivers in some way.

Of course, redeveloping waterfronts has plenty of downsides. By definition, river floodplains flood, so building on them always carries risk. Climate change is making flood probabilities harder to predict, and, in many areas, the risks are expected to worsen.

In coastal areas, storm surges and long-term sea level rise pose existential threats to waterfront properties. Meanwhile, heavy industries displaced from affluent cities are spreading in developing countries and into some of the world's last remaining pristine areas, creating new problems of river pollution, water diversions, and ecological destruction elsewhere.

Other issues are purely socioeconomic. In unaffordable cities like London, Shanghai, and Los Angeles, replacing dilapidated industrial corridors with gleaming new waterfront developments raises real estate prices in the surrounding neighborhoods, displacing low- and middle-income residents. Every lost affordable neighborhood reduces a city's overall social and economic diversity, chipping away at the very culture and vibrancy of the metropolis itself.

Urban riverfronts are typically public spaces used for many purposes, including unauthorized street art. This interesting example appeared under a bridge spanning the Danube River in Vienna, Austria, in 2017. *(Mia Bennett)*

Urban riverfront renewals also harm other groups who use them as a sort of public commons. These include homeless encampments, some quite large, like those established along the Rio Negro River in Manaus, Brazil, and the Santa Ana River in Orange County, California. They also include street artists, who use bridge abutments and cement-lined channels as blank canvases for their craft.

From Greenpoint, Brooklyn, to Maspero Triangle, Cairo, plenty of unhappy locals will attest that these downsides are not halting the urban riverfront renewal boom. Most cities are growing and their cores are densifying. More young people and working professionals are choosing downtowns over suburbia. Large, well-located tracts of formerly industrial land, lost

Riverfront festivals are held regularly in downtown Providence, Rhode Island, attracting thousands to an outdoor experience. They signify a dramatic return of humans to rivers of the Blackstone River Valley, once the heart of America's Industrial Revolution and one of the most polluted waterways in America. *(Abbie Tingstad)*

dockyards, and petrochemical plants are being replaced by new parks, residential towers, and service-sector jobs. Housing and greenspace are scarce, and city planners are rediscovering their riverfronts to encourage both.

On balance, this is a good thing. Done right, urban riverfront renewals offer a rare opportunity to create dense, appealing neighborhoods with access to a calm outdoor setting and a curated form of nature. Already, millions of city people are enjoying newly built riverfront parks, where they can stroll, exercise, or simply pass a few mildly stimulating minutes outside. In cities large and small, these parks create public commons where people assemble for cultural events, such as the ongoing WaterFire festivals that attract nearly a million people each summer to downtown Providence, Rhode Island, once one of the most polluted river corridors in America.

One doesn't have to hike the Appalachian Trail to reap the cognitive and health benefits of outdoor settings. Riverfront spaces will make this a daily possibility for millions—yes, billions—of people in our rapidly urbanizing world.

Rivers of Power

A world without rivers would be unrecognizable to us. Our continents would be rugged, high, cold, and small. Our settlement pattern would have evolved in very different ways, with scattered farms and villages clinging to oases and coastlines. Wars would have proceeded differently, and the borders of nations would be unfamiliar. Our most famous cities would not exist. The global movements of people and trade that so define us today might never have happened.

Throughout history, rivers have gripped us through their provisions of natural capital, access, territory, well-being, and power. The flat, fertile valleys they created sustained us with food and water. From the silty valleys of present-day Egypt, Iraq, India, Pakistan, and China, the first hydraulic societies invented the city, trade, the ruling class, and the origins of the political state. The pragmatic desire of early urban planners to tap water and flush sewage fashioned the world's first engineers. Philosophical debates over the sources and ownership of rivers established important early foundations of science and law.

In the Americas, prehistoric colonizers spread out along river valleys and built other hydraulic societies, like the Anasazi of New Mexico, the Maya of Central America, and the Cahokians of the Mississippi River valley. When colonial Europeans set out to chart the world, they used rivers as exploration corridors and their courses and topographic divides to define territorial claims. Many still persist as the borders of national and subnational political jurisdictions today. We dotted the world's riverbanks with settlements that became towns, cities, and eventually metropolises.

Successive generations depended on rivers without giving them much thought. They were just there, a pleasant landscape feature with some essential but narrow value. Provider of fish. Irrigator of hydraulic kingdoms. Pathways to explore continents. Enabler of industrialization. Flusher of poisons. Maker of electricity. Opener of arid lands. Coolant of power plants. Inspirer of environmental and technological movements. Opportunity for real estate development. Soother of stressed urban minds. Viewed from any one generation, the value of rivers is obvious, utilitarian, even banal. Only by taking the long view is their foundational importance to human civilization revealed.

If I were to condense this long history of rivers and people into a tidy story for my three children, it might go something like this:

Once upon a time, the rains came together and shaped the land.

For millions of years, flowing water carved the mountains and carried the earth. Rivers grew the land out into the sea and built wide valley plains of rich alluvial soil.

Nomadic humans found these valleys and learned to farm, cooperate, and live in one place. Rivers became essential to their way of life. Food surpluses increased, and societies became more complex and hierarchical. They began to support thinkers who wondered about the natural world. The first glimmerings of philosophy, law, engineering, and science emerged. We learned and traded and became more creative and more populous.

Our use of rivers expanded. They became corridors of travel and, eventually, a means to re-explore the continents. We founded new settlements, and our cultures and languages diversified. As our numbers grew, so also did the death tolls, as floods intermittently killed us and jolted our political systems in unpredictable ways. Rivers became stratagems of warfare and convenient arbiters of territory in an increasingly competitive world.

Our technologies advanced and we learned to press rivers into industry. We built waterwheels, tanneries, and textile mills. Rivers lent their power and flushed away poisons and helped industrialize our economy. They floated barges and trade ships, and we dug artificial rivers—canals—to interlink them better.

Then, our technological prowess leaped. We learned to subjugate rivers at massive scale, at a price of physically driving us away from them. By swallowing rivers in huge reservoirs, we settled arid lands previously thought unlivable. Like cornucopias, the big dams gave us water, electricity, and towns. Their waters became too valuable to risk losing, leading to extraordinary new treaties for cooperation and shared governance of transboundary rivers.

As our wealth grew, our tolerance of filth weakened. Polluted waterways, once an acceptable casualty of economic growth, became outrageous. Laws were changed and poisons curbed. The rivers rebounded and we learned of their great elasticity. Dams were torn down and the rivers answered, swiftly resuming their transport of sediment from land to sea as fish returned home past the blown abutments.

The story is now in the present. We possess godlike earth-moving power and the engineering knowledge to take river diversions to an extraordinary scale. The world's two most populous countries, facing daunting water-supply challenges, are advancing the biggest interbasin water transfer schemes ever conceived. The ecological health of rivers, while still a priority, is under pressure. We are building new sensors, satellites, and models to better understand them and monitor their state. New technologies may tap river power in less environmentally damaging ways.

Meanwhile, we are moving into great cities, immersed in the digital world, the natural one fading from our everyday experience. But the human brain and physiology have not kept up. For our own well-being, some benefits of preserving contact with nature remain. Once again the rivers, around which our cities are hardening, are there.

They have always been there, in essential and evolving ways, every step of the way. Like the child growing up in Shel Silverstein's *The Giving Tree,* our demands changed over time and the rivers' gifts changed with us. Our needs evolved, but the dependency remained. Rivers have been propping us up for hundreds of generations. And unlike Silverstein's withered apple tree stump, they will rejuvenate if we let them. Rivers can be immortal.

A Nilometer, that ancient source of knowledge once used by pharaohs to declare taxes and preserve their society, lies just downstream of the chosen site of Cairo's Nile Tower. The Nile has always provided gifts to Egyptians, first silt-rich farmland

and irrigating floodwaters, then energy from the Aswan High Dam, and now upscale real estate along the downtown Cairo riverfront.

From the invention of the city-state to the exploration of our planet, from the contesting of territory to the birth of cities, from the capture of energy to the industrialization of economies, from motivating cooperation, environmentalism, and technology to serving up curated pockets of nature for billions of urban people, the rivers are there.

There is a vast, arterial power humming all around us, hiding in plain sight. It has shaped our civilization more than any road, technology, or political leader. It has opened frontiers, founded cities, settled borders, and fed billions. It promotes life, forges peace, grants power, and capriciously destroys everything in its path. Increasingly domesticated, even manacled, it is an ancient force that rules us still.

ACKNOWLEDGMENTS

I am grateful to my parents, for making this book possible. My father, Norman D. Smith, introduced me to rivers at multiple stages in my life, first as the parent of an urban child and later through his science, as an eminent fluvial sedimentologist and authoritative expert on Canada's Saskatchewan River. Thank you.

This book would never have been written without the encouragement and high standards of literary agent Russell Weinberger of Brockman, Inc., who has stuck with me since the beginning and insists that I write only about things I know.

I thank Ian Straus, associate editor at Little, Brown, who gave magnanimously of his time, including multiple editorial reviews of the full manuscript. Tracy Behar, vice president, publisher, and editor-in-chief, kindly agreed to publish the book. Betsy Uhrig, Kathryn Rogers, Jessica Chun, and Juliana Horbachevsky respectively led the book's production editing, copyediting, marketing, and publicity. Striking cover art for the hardcover edition was designed by Lauren Harms.

Two former undergraduate students, Natalie Pearl from Brown University and Sarah Popelka from the University of California, Los Angeles, provided invaluable research assistance. All maps and illustrations were created in collaboration with the superb cartographer Matthew Zebrowski, in the UCLA Department of Geography.

Financial support for some elements of this book was made possible in part through grants from Brown University (Institute at Brown for Environment & Society), John Atwater and Diana Nelson, and the National Aeronautics and Space Administration (NASA) Earth Science Division.

Individuals who provided interviews, readings, research support, advice, or other helpful forms of influence include Fred Adjarian, John Agnew, Tesfay Alemseged, Doug Alsdorf, Kostas Andreadis, Lorena Apodaco, Gedion Asfaw, Paul Bates, Alberto Behar, Jason Box, Rachel Calico, Caitlin Campbell, Judy Carney, William A. V. Clark, Adrian Clayton, Kyli Cosper, John Crilley, Angela DeSoto, Jared Diamond, Mike Durand, Corey Eide, Jared Entin, Jay Famiglietti, Wubalem Fekade, James Garvin, Mekonnen Gebremichael, Pam Giesel, Tom Gillespie, Peter Griffith, Colene Haffke, Tyler Harlan, Line Haug, Jessy Jenkins, Chris Johnson, Yara Khoshnaw, Jeffrey Kightlinger, William Krabill, Yumiko Kura, Scott LeFavour, Carl Legleiter, Dennis Lettenmaier, Adam LeWinter, Eric Lindstrom, Richard Lorman, Lula Lu, Amanda Lynch, Glen MacDonald, Frank Magilligan, Hank Margolis, Thorsten Markus, Kasi McMurray, Frode Mellemvik, Leal Mertes, Charles Miller, Cory Milone, Toby Minear, Nicole Morales, Paul Morin, Irene Mortensen, Becky Mudd, Fekahmed Negash, Petter Norre, Larry Nulty, Greg Okin, Brandon Overstreet, Fred Pearce, Al Pietroniro, Erica Pietroniro, Try Pisey, Bob Pries, Liz Pulver, Wesley Reisser, Jovana Rizzo, Ernesto Rodriguez, Sok Sovanary, Joanne Stokes, Karen Tamir, Marco Tedesco, Arja Tingstad, Jerry Tingstad, Dirk van As, Sophirun Ven, Thomas Wagner, Jida Wang, Michael Wehner, Cindy Ye, and Kathy Young.

Knowingly or not, many of my graduate students and postdoctoral scholars influenced this book: Mia Bennett, Vena Chu, Sarah Cooley, Matthew Cooper, Jessica Fayne, Karen Frey, Colin Gleason, Cynthia Hall, Ethan Kyzivat, Ekaterina Lezine, Matthew Mersel, Tamlin Pavelsky, Lincoln Pitcher, Åsa Rennermalm, John Ryan, Yongwei Sheng, Scott Stephenson, and Kang Yang.

Non-author photographs were provided by Gedion Asfaw, Mia Bennett, John Gussman, Tyler Harlan, Michal Huniewicz, Richard Lorman, Mann Power Hydro Ltd/David Mann, the National Aeronautics and Space Administration, the Library of Congress, the Richard Nixon Presidential Library and Museum, John Ryan, Abbie Tingstad, Kelvin Trautman, and the United Nations Refugee Agency.

An initial draft of this book was written at Brown University during a yearlong sabbatical supported by the University of California, Los Angeles. The manuscript was subsequently greatly improved thanks to critical reviews of key chapters by Tamlin Pavelsky, Jerry Tingstad, and Kang Yang. Doug Alsdorf, Norman Smith, Ian Straus, and Abbie Tingstad read and critiqued the book in its entirety. Professional fact-checking was provided by Sarah Lippincott. Responsibility for any remaining errors, inaccuracies, or omissions is of course my own.

Laurence C. Smith
Providence, Rhode Island, USA
2 December 2019

REFERENCES AND FURTHER READING

Introduction

Holden, Peter, et al. "Mass-spectrometric mining of Hadean zircons by automated SHRIMP multi-collector and single-collector U/Pb zircon age dating: The first 100,000 grains," *International Journal of Mass Spectrometry* 286.2–3 (2009): 53–63. doi.org/10.1016/j.ijms.2009.06.007

Kite, Edwin S., et al. "Persistence of intense, climate-driven runoff late in Mars history," *Science Advances* 5.3 (2019). doi.org/10.1126/sciadv.aav7710

Lyons, Timothy W., et al. "The rise of oxygen in Earth's early ocean and atmosphere," *Nature* 506 (2014): 307–315. doi.org/10.1038/nature13068

O'Malley-James, Jack T., et al. "Swansong Biospheres: Refuges for Life and Novel Microbial Biospheres on Terrestrial Planets near the End of Their Habitable Lifetimes," *International Journal of Astrobiology* 12.2 (2012): 99–112. doi.org/10.1017/S147355041200047X

Valley, John W., et al. "Hadean age for a post-magma-ocean zircon confirmed by atom-probe tomography," *Nature Geoscience* 7 (2014): 219–223. doi.org/10.1038/ngeo2075

Chapter 1 — The Palermo Stone

The Land Between the Rivers

Fagan, Brian. *Elixir: A History of Water and Humankind* (New York: Bloomsbury Press, 2011).

Hurst, H. E. "The Roda Nilometer." (Book Review of *Le Mikyâs ou Nilometrè de l'lle de Rodah, Par Kamel Osman Ghaleb Pasha.*) *Nature* 170 (1952): 132–133. doi.org/10.1038/170132a0

Morozova, Galina S. "A review of Holocene avulsions of the Tigris and Euphrates rivers and possible effects on the evolution of civilizations in lower Mesopotamia," *Geoarchaeology* 20.4 (2005): Wiley Online Library. doi.org/10.1002/gea.20057

Shaw, Ian, ed. *The Oxford History of Ancient Egypt* (New York: Oxford University Press, 2000).

Tainter, Joseph A. *The Collapse of Complex Societies* (Cambridge: Cambridge University Press, 1990).

Ark of the Tigris-Euphrates?

Davila, James R. "The Flood Hero as King and Priest," *Journal of Near Eastern Studies* 54.3 (1995): 199–214.

Kennett, D. J., and J. P. Kennett. "Early State Formation in Southern Mesopotamia: Sea Levels, Shorelines, and Climate Change," *Journal of Island and Coastal Archaeology* 1.1 (2006): 67–99. doi.org/10.1080/15564890600586283

Lambeck, K. "Shoreline reconstructions for the Persian Gulf since the last glacial maximum," *Earth and Planetary Science Letters* 142.1–2 (1996): 43–57. doi.org/10.1016/0012-821X(96)00069-6

Lambeck, K., and J. Chappell. "Sea Level Change Through the Last Glacial Cycle," *Science* 292.5517 (2001): 679–686. doi.org/10.1126/science.1059549

Ryan, W. B. F., et al. "Catastrophic Flooding of the Black Sea," *Annual Review of Earth and Planetary Sciences* 31 (2003): 525–554. doi.org/10.1146/annurev.earth.31.100901.141249

Teller, J. T., et al. "Calcareous dunes of the United Arab Emirates and Noah's Flood: the postglacial reflooding of the Persian (Arabian) Gulf," *Quaternary International* 68–71 (2000): 297–308. doi.org/10.1016/S1040-6182(00)00052-5

Tigay, Jeffrey H. *The Evolution of the Gilgamesh Epic* (Philadelphia: University of Pennsylvania Press, 1982).

Secrets of the Sarasvati

Gangal, K., et al. "Spatio-temporal analysis of the Indus urbanization," *Current Science* 98.6 (2010): 846–852. www.jstor.org/stable/24109857

Giosan, L., et al. "Fluvial landscapes of the Harappan civilization," *PNAS* 109.26 (2012): 1688–1694. doi.org/10.1073/pnas.1112743109

Sarkar, A., et al. "Oxygen isotope in archaeological bioapatites from India: Implications to climate change and decline of Bronze Age Harappan civilization," *Scientific Reports* 6 (2016). doi.org/10.1038/srep26555

Tripathi, J. K., et al. "Is River Ghaggar, Saraswati? Geochemical Constraints," *Current Science* 87.8 (2004): 1141–1145. www.jstor.org/stable/24108988

Yu the Great Returns; Wittfogel's Waterworld; Knowledge, from the Breasts of Hapi

Biswas, A. K. *History of Hydrology* (London: North-Holland Publishing Company, 1970).

Loewe, Michael, and Edward L. Shaughnessy, eds. *The Cambridge History of Ancient China: From the Origins of Civilization to 221 BC* (Cambridge: Cambridge University Press, 1999).

Makibayashi, K., "The Transformation of Farming Cultural Landscapes in the Neolithic Yangtze Area, China," *Journal of World Prehistory* 27.3–4 (2014): 295–307. doi.org/10.1007/s10963-014-9082-0

Mays, Larry W. "Water Technology in Ancient Egypt," *Ancient Water Technologies* (Dordrecht, Netherlands: Springer, 2010).

Truesdell, W. A. "The First Engineer," *Journal of the Association of Engineering Societies* 19 (1897): 1–19.

Wittfogel, Karl A. *Oriental Despotism: A Comparative Study of Total Power* (New Haven: Yale University Press, 1957).

Wu, Q., et al. "Outburst flood at 1920 BCE supports historicity of China's Great Flood and the Xia dynasty," *Science* 353.6299 (2016): 579–582. doi.org/10.1126/science.aaf0842

Zong, Y., et al. "Fire and flood management of coastal swamp enabled first rice paddy cultivation in east China," *Nature* 449.7161 (2007): 459–462. doi.org/10.1038/nature06135

The Hammurabi Code; Rivers for All

Bannon, Cynthia. "Fresh Water in Roman Law: Rights and Policy," *Journal of Roman Studies* 107 (2017): 60–89. doi.org/10.1017/S007543581700079X

Campbell, Brian. *Rivers and the Power of Ancient Rome* (Chapel Hill: University of North Carolina Press, 2012).

Finkelstein, J. J. "The laws of Ur-Nammu," *Journal of Cuneiform Studies* 22.3–4 (1968): 66–82. doi.org/10.2307/1359121

——— "Sex Offenses in Sumerian Laws," *Journal of the American Oriental Society* 86.4 (1966): 355–372. doi.org/10.2307/596493

Frymer, T. S. "The Nungal-Hymn and the Ekur-Prison," *Journal of the Economic and Social History of the Orient* 20.1 (1977): 78–89. doi.org/10.2307/3632051

Gomila, M. "Ancient Legal Traditions," *The Encyclopedia of Criminology and Criminal Justice* (2014): 1–7. Wiley Online Library. doi.org/10.1002/9781118517383.wbeccj252

Husain, M. Z., and S. E. Costanza. "Code of Hammurabi," *The Encyclopedia of Corrections* (2017): 1–4. Wiley Online Library. doi.org/10.1002/9781118845387.wbeoc034

Teclaff, Ludwik A. "Evolution of the River Basin Concept in National and International Water Law," *Natural Resources Journal* 36.2 (1996): 359–391. digitalrepository.unm.edu/nrj/vol36/iss2/7

Yildiz, F. "A Tablet of Codex Ur-Nammu from Sippar," *Orientalia* 50.1 (1981): 87–97. www.jstor.org/stable/43075013

Wheels of Power; Valleys of the New World; George Washington's Big America

Arnold, Jeanne E. "Credit Where Credit Is Due: The History of the Chumash Oceangoing Plank Canoe," *American Antiquity* 72.2 (2007): 196–209. doi.org/10.2307/40035811

Canuto, Marcello A., et al. "Ancient lowland Maya complexity as revealed by airborne laser scanning of northern Guatemala," *Science* 361.6409 (2018): doi.org/10.1126/science.aau0137

Cleland, Hugh. *George Washington in the Ohio Valley* (Pittsburgh: University of Pittsburgh Press, 1955).

"The Founders and the Pursuit of Land," The Lehrman Institute, lehrmaninstitute.org/history/founders-land.html#washington

Davis, Loren G., et al. "Late Upper Paleolithic occupation at Cooper's Ferry,

Idaho, USA, ~16,000 years ago," *Science* 365.6456 (2019): 891–897. doi: 10.1126/science.aax9830

Liu, Li, and Leping Jiang. "The discovery of an 8000-year-old dugout canoe at Kuahuqiao in the Lower Yangzi River, China," *Antiquity* 79.305 (2005): www.antiquity.ac.uk/projgall/liu305/

Pauketat, T. R. *Ancient Cahokia and the Mississippians* (New York: Cambridge University Press, 2004).

Pepperell, Caitlin S., et al. "Dispersal of *Mycobacterium tuberculosis* via the Canadian fur trade," *Proceedings of the National Academy of Sciences (PNAS)* 108.16 (2011): 6526–6531. doi.org/10.1073/pnas.1016708108

Van de Noort, R., et al. "The 'Kilnsea-boat,' and some implications from the discovery of England's oldest plank boat remains," *Antiquity* 73.279 (1999): 131–135. doi.org/10.1017/S0003598X00087913

Wade, L. "Ancient site in Idaho implies first Americans came by sea," *Science* 365.6456 (2019): 848–849. doi: 10.1126/science.365.6456.848

Chapter 2 — On the Border

Blue Borders; Lines of Expediency

Apodaca, Lorena (Border Patrol Agent, Border Community Liaison, U.S. Customs and Border Protection) and Irine Mortenson (Community Relations Officer, U.S. Customs and Border Protection). Personal interviews. 14 Aug. 2017. El Paso, Texas.

Carter, Claire, et al. *David Taylor: Monuments* (Radius Books/Nevada Museum of Art, 2015).

Popelka, Sarah J., and Laurence C. Smith. "Rivers as Political Borders: A New Subnational Geospatial Dataset," *Water Policy*, in review.

Reisser, Wesley J. *The Black Book: Woodrow Wilson's Secret Plan for Peace* (Lanham, MD: Lexington Books, 2012).

Sahlins, Peter. "Natural Frontiers Revisited: France's Boundaries since the Seventeenth Century," *The American Historical Review* 95.5 (1990): 1423–1451. doi.org/10.2307/2162692

Ullah, Akm Ahsan. "Rohingya Refugees to Bangladesh: Historical Exclusions and Contemporary Marginalization," *Journal of Immigrant & Refugee Studies* 9.2 (2011): 139–161. doi.org/10.1080/15562948.2011.567149

The Size and Shape of Nations; Worries of Water Wars; Mandela the Bomber; Water Towers Make Water Wards

Alesina, Alberto, and Enrico Spolaore. *The Size of Nations* (Cambridge: MIT Press, 2005).

Likoti, Fako Johnson. "The 1998 Military Intervention in Lesotho: SADC Peace Mission or Resource War?" *International Peacekeeping* 14.2 (2007): 251–263. doi.org/10.1080/13533310601150875

Makoa, Francis K. "Foreign military intervention in Lesotho's election dispute: Whose Project?" *Strategic Review for Southern Africa* 21.1 (1999).

Viviroli, Daniel, et al. "Mountains of the world, water towers for humanity:

Typology, mapping, and global significance," *Water Resources Research* 43.7 (2007): 1–13. doi.org/10.1029/2006WR005653

Harmon's Folly; All Eyes on the Mekong

Convention between the United States of America and Mexico: Equitable Distribution of the Waters of the Rio Grande. Proclaimed January 16, 1907, by the U.S. and Mexico. www.ibwc.gov/Files/1906Conv.pdf

Convention on the Law of the Non-Navigational Use of International Watercourses, United Nations. Adopted 1997. legal.un.org/avl/ha/clnuiw/clnuiw.html

Cosslett, Tuyet L., and Patrick D. Cosslett. *Sustainable Development of Rice and Water Resources in Mainland Southeast Asia and Mekong River Basin* (Singapore: Springer Nature, 2018).

Harris, Maureen. "Can regional cooperation secure the Mekong's future?" *Bangkok Post,* 10 January 2018. www.bangkokpost.com/opinion/opinion/1393266/can-regional-cooperation-secure-the-mekongs-future

McCaffery, Stephen C. "The Harmon Doctrine One Hundred Years Later: Buried, Not Praised," *Natural Resources Journal* 36.3 (1996): 549–590. digitalrepository.unm.edu/nrj/vol36/iss3/5

Middleton, Carl, and Jeremy Allouche. "Watershed or Powershed? Critical Hydropolitics, China and the Lancang-Mekong Cooperation Framework," *The International Spectator* 51.3 (2016): 100–117. doi.org/10.1080/03932729.2016.1209385

Salman, Salman M. A. "Entry into force of the UN Watercourses Convention: Why should it matter?" *International Journal of Water Resources Development* 31.1 (2015): 4–16. doi.org/10.1080/07900627.2014.952072

Schiff, Jennifer S. "The evolution of Rhine river governance: historical lessons for modern transboundary water management," *Water History* 9.3 (2017): 279–294. doi.org/10.1007/s12685-017-0192-3

Sosland, Jeffrey K. *Cooperating Rivals: The Riparian Politics of the Jordan River Basin* (Albany: State University of New York Press, 2007).

Teclaff, Ludwik A. "Fiat or Custom: The Checkered Development of International Water Law," *Natural Resources Journal* 31.1 (1991): 45–73. digitalrepository.unm.edu/nrj/vol31/iss1/4

Wolf, Aaron T. "Conflict and cooperation along international waterways," *Water Policy* 1.2 (1998): 251–265. doi.org/10.1016/S1366-7017(98)00019-1

Ziv, Guy, et al. "Trading-off fish biodiversity, food security, and hydropower in the Mekong River Basin," *PNAS* 109.15 (2012): 5609–5614. doi.org/10.1073/pnas.1201423109

Chapter 3 — The Century of Humiliation and Other War Stories

Cross That River

Brockell, Gillian. "How a painting of George Washington crossing the Delaware on Christmas went 19th-century viral," *The Washington Post,* 25 Dec. 2017. www

.washingtonpost.com/news/retropolis/wp/2017/12/24/how-a-painting-of-george-washington-crossing-the-delaware-on-christmas-went-19th-century-viral/?utm_term=.05b8375ce759

Groseclose, Barbara S. "'Washington Crossing the Delaware': The Political Context," *The American Art Journal* 7.2 (1975): 70–78. www.jstor.org/stable/1594000

"Islamic State and the crisis in Iraq and Syria in maps," *BBC News*, 28 Mar. 2018. www.bbc.com/news/world-middle-east-27838034

Jones, Seth G., et al. *Rolling Back the Islamic State* (Santa Monica, CA: RAND Corporation, 2017). www.rand.org/pubs/research_reports/RR1912.html

America Divided

Bearss, Edwin C., with J. Parker Hills. *Receding Tide: Vicksburg and Gettysburg: The Campaigns That Changed the Civil War* (Washington, DC: National Geographic Books, 2010).

Joiner, Gary D. *Mr. Lincoln's Brown Water Navy: The Mississippi Squadron* (Lanham, MD: Rowman & Littlefield Publishers, 2007).

Tomblin, Barbara Brooks. *The Civil War on the Mississippi: Union Sailors, Gunboat Captains, and the Campaign to Control the River* (Lexington: University Press of Kentucky, 2016).

Van Tilburg, Hans Konrad. *A Civil War Gunboat in Pacific Waters: Life on Board USS Saginaw* (Gainesville: University Press of Florida, 2010).

The Century of Humiliation

Chang, Iris. *The Rape of Nanking: The Forgotten Holocaust of World War II* (New York: Basic Books, Reprint Edition, 2012).

Cole, Bernard D. "The Real Sand Pebbles," *Naval History Magazine* 14.1 (2000): U.S. Naval Institute. www.usni.org/magazines/naval-history-magazine/2000/february

Feige, Chris, and Jeffrey A. Miron. "The opium wars, opium legalization and opium consumption in China," *Applied Economics Letters* 15.12 (2008): 911–913. doi.org/10.1080/13504850600972295

"A Japanese Attack Before Pearl Harbor," *NPR Morning Edition*, 13 Dec. 2007. Audio here: www.npr.org/templates/story/story.php?storyId=17110447

Kaufman, Alison A. *The "Century of Humiliation" and China's National Narratives* (2011). www.uscc.gov/sites/default/files/3.10.11Kaufman.pdf

Konstam, Angus. *Yangtze River Gunboats 1900–49* (Oxford, U.K.: Osprey Publishing, 2011).

Melancon, Glenn. "Honour in Opium? The British Declaration of War on China, 1839–1840," *The International History Review* 21.4 (1999): 855–874. doi.org/10.1080/07075332.1999.9640880

"The Opening to China Part 1: the First Opium War, the United States, and the Treaty of Wangxia, 1839–1844." Office of the Historian, U.S. Department of State. history.state.gov/milestones/1830-1860/china-1

"The Opening to China Part 2: the Second Opium War, the United States, and the Treaty of Tianjin, 1857–1859." Office of the Historian, U.S. Department of State. history.state.gov/milestones/1830-1860/china-2

"USS *Saginaw,*" military.wikia.com/wiki/USS_Saginaw

Rivers of Metal; Britain's Dambusters; The Matador's Cloak

Builder, Carl H., et al. "The Technician: Guderian's Breakthrough at Sedan," *Command Concepts,* 43–54 (Santa Monica, CA: RAND Corporation, 1999). www.rand.org/content/dam/rand/pubs/monograph_reports/MR775/MR775.chap4.pdf

Cole, Hugh M. "The Battle Before the Meuse," *The Ardennes: Battle of the Bulge.* U.S. Army Center of Military History. history.army.mil/books/wwii/7-8/7-8_22.HTM

"The Dam Raids," The Dambusters. http://www.dambusters.org.uk/

Evenden, Matthew. "Aluminum, Commodity Chains, and the Environmental History of the Second World War," *Environmental History* 16.1 (2011): 69–93. doi.org/10.1093/envhis/emq145

Hall, Allan. "Revealed: The priest who changed the course of history... by rescuing a drowning four-year-old Hitler from death in an icy river," *Mailonline,* 5 Jan. 2012. www.dailymail.co.uk/news/article-2082640/How-year-old-Adolf-Hitler-saved-certain-death—drowning-icy-river-rescued.html

King, W. L. Mackenzie. "The Hyde Park Declaration," *Canada and the War.* 28 Apr. 1941. wartimecanada.ca/sites/default/files/documents/WLMK.HydePark.1941.pdf

Massell, David. "'As Though There Was No Boundary': the Shipshaw Project and Continental Integration," *American Review of Canadian Studies* 34.2 (2004): 187–222. doi.org/10.1080/02722010409481198

Royal Air Force Benevolent Fund. "The story of the Dambusters." www.rafbf.org/dambusters/the-story-of-the-dambusters

Webster, T. M. "The Dam Busters Raid: Success or Sideshow?" *Air Power History* 52.2 (2005): 12–25. www.questia.com/library/journal/1G1-133367811/the-dam-busters-raid-success-or-sideshow

Willis, Amy. "Adolf Hitler 'Nearly Drowned as a Child,'" *The Telegraph,* 6 Jan. 2012. www.telegraph.co.uk/history/world-war-two/8996576/Adolf-Hitler-nearly-drowned-as-a-child.html

Zahniser, Marvin R. "Rethinking the Significance of Disaster: The United States and the Fall of France in 1940," *International History Review* 14.2 (1992): 252–276. www.jstor.org/stable/40792747

A Milk Run in Vietnam

Carrico, John M. *Vietnam Ironclads: A Pictorial History of U.S. Navy River Assault Craft, 1966–1970* (John M. Carrico, 2008).

Dunigan, Molly, et al. *Characterizing and exploring the implications of maritime irregular warfare.* RAND Document Number MG-1127-NAVY (Santa Monica, CA: RAND, 2012) www.rand.org/pubs/monographs/MG1127.html

Helm, Glenn E. "Surprised at TET: U.S. Naval Forces—1968," Mobile Riverine Force Association. www.mrfa.org/us-navy/surprised-at-tet-u-s-naval-forces-1968/

Lorman, Richard E. "The Milk Run," *River Currents* 23.4 (2014): 2–3. Mobile Riverine Force Association. www.mrfa.org/wp-content/uploads/2016/06/RiverCurrentsWinter2014WEB.pdf

———. Personal interview. 22 April 2018. Hull, Massachusetts.

Qiang, Zhai. "China and the Geneva Conference of 1954," *China Quarterly* 129 (1992): 103–122. www.jstor.org/stable/654599

Chapter 4 — Ruin and Renewal

After the Deluge

"Billion-Dollar Weather and Climate Disasters: Overview." NOAA National Centers for Environmental Information. www.ncdc.noaa.gov/billions/

Blake, Eric S., and David A. Zelinsky. *National Hurricane Center Tropical Cyclone Report: Hurricane Harvey* (NOAA, 2018). www.nhc.noaa.gov/data/tcr/AL092017_Harvey.pdf

Elliott, James R. "Natural Hazards and Residential Mobility: General Patterns and Racially Unequal Outcomes in the United States," *Social Forces* 93.4 (2015): 1723–1747. doi.org/10.1093/sf/sou120

Fussell, Elizabeth, et al. "Race, Socioeconomic Status, and Return Migration to New Orleans after Hurricane Katrina," *Population and Environment* 31.1–3 (2010): 20–42. doi.org/10.1007/s11111-009-0092-2

Haider-Markel, Donald P., et al. "Media Framing and Racial Attitudes in the Aftermath of Katrina," *Policy Studies Journal* 35.4 (2007): 587–605. doi.org/10.1111/j.1541-0072.2007.00238.x

"Hurricane Costs." NOAA Office for Coastal Management. www.coast.noaa.gov/states/fast-facts/hurricane-costs.html

Jonkman, Sebastian N., et al. "Brief communication: Loss of life due to Hurricane Harvey," *Natural Hazards and Earth System Sciences* 18.4 (2018): 1073–1078. doi.org/10.5194/nhess-18-1073-2018

Leeson, Peter T., and Russell S. Sobel. "Weathering Corruption," *Journal of Law and Economics* 51.4 (2008): 667–681. doi.org/10.1086/590129

Schultz, Jessica, and James R. Elliott. "Natural disasters and local demographic change in the United States," *Population and Environment* 34.3 (2013): 293–312 www.jstor.org/stable/42636673

SHELDUS (Spatial Hazard Events and Losses Database for the United States), University of South Carolina. cemhs.asu.edu/sheldus/

Team Rubicon (various volunteers). Personal interviews. 29–30 Sept. 2017. Houston and Conroe, Texas.

Zaninetti, Jean-Marc, and Craig E. Colten. "Shrinking New Orleans: Post-Katrina Population Adjustments," *Urban Geography* 33.5 (2012): 675–699. doi.org/10.2747/0272-3638.33.5.675

When the Levees Break

Barry, John S. *Rising Tide: The Great Mississippi Flood of 1927 and How It Changed America* (New York: Simon & Schuster, 1998).

Rivera, Jason David, and DeMond Shondell Miller. "Continually Neglected: Situating Natural Disasters in the African American Experience," *Journal of Black Studies* 37.4 (2007): 502–522. www.jstor.org/stable/40034320

Walton Jr., Hanes, and C. Vernon Gray. "Black Politics at the National Republican and Democratic Conventions, 1868–1972," *Phylon* 36.3 (1975): 269–278. www.jstor.org/stable/274392

Weaponizing China's Sorrow

Alexander, Bevin. *The Triumph of China*. www.bevinalexander.com/china/

Edgerton-Tarpley, Kathryn. "A River Runs through It: The Yellow River and The Chinese Civil War, 1946–1947," *Social Science History* 41.2 (2017): 141–173. doi.org/10.1017/ssh.2017.2

———. "'Nourish the People' to 'Sacrifice for the Nation': Changing Responses to Disaster in Late Imperial and Modern China," *Journal of Asian Studies* 73.2 (2014): 447–469. www.jstor.com/stable/43553296

Lary, Diana. "Drowned Earth: The Strategic Breaching of the Yellow River Dyke, 1938," *War in History* 8.2 (2001): 191–207. doi.org/10.1177/096834450100800204

Muscolino, Micah S. "Refugees, Land Reclamation, and Militarized Landscapes in Wartime China: Huanglongshan, Shaanxi, 1937–45," *Journal of Asian Studies* 69.2 (2010): 453–478. www.jstor.org/stable20721849

———. "Violence Against People and the Land: The Environment and Refugee Migration from China's Henan Province, 1938–1945," *Environment and History* 17.2 (2011): 291–311. www.jstor.org/stable/41303510

Muscolino, Micah. "War, Water, Power: An Environmental History of Henan's Yellow River Flood Area, 1938–1952," *CEAS Colloquium Series*. 9 Apr. 2012. Yale Macmillan Center.

Phillips, Steven E. *Between Assimilation and Independence: the Taiwanese Encounter Nationalist China, 1945–1950* (Redwood City, CA: Stanford University Press, 2003).

Rubinstein, Murray A., ed. *Taiwan: A New History* (Armonk, NY: M.E. Sharpe, 1999).

Selden, Mark, and Alvin Y. So. *War & State Terrorism: The United States, Japan, and the Asia-Pacific in the Long Twentieth Century* (Lanham, MD: Rowman & Littlefield, 2004).

Shu, Li, and Brian Finlayson. "Flood management on the lower Yellow River: hydrological and geomorphological perspectives," *Sedimentary Geology* 85.1–4 (1993): 285–296. doi.org/10.1016/0037-0738(93)90089-N

It Came from the South Fork Club

Connelly, Frank, and George C. Jenks. *Official History of the Johnstown Flood (1889)*, (Pittsburgh: Journalist Publishing Company, 1889).

Eaton, Lucien, et al. *The American Law Review, Volume 23* (St. Louis, MO: Review Publishing Co., 1889), 647.

Shugerman, Jed H. "The Floodgates of Strict Liability: Bursting Reservoirs and the Adoption of *Fletcher v. Rylands* in the Gilded Age," *Yale Law Journal,* 110.2 (2000). digitalcommons. law.yale.edu/ylj/vol110/iss2/6

Simpson, A. W. B. "Legal Liability for Bursting Reservoirs: The Historical Context of 'Rylands v. Fletcher,'" *Journal of Legal Studies,* 13.2 (1984): 209–264. www.jstor.org/stable/724235

Chapter 5 — Seizing the Current

The GERD; The Megadam Century

Abdalla, I. H. "The 1959 Nile Waters Agreement in Sudanese-Egyptian Relations," *Middle Eastern Studies* 7.3 (1971): 329–341. www.jstor.org/stable/4282387

Bochove, Danielle, et al. "Barrick to Buy Randgold to Expand World's Largest Gold Miner," *Bloomberg,* 24 Sept. 2018. www.bloomberg.com/news/articles/2018-09-24/barrick-gold-agrees-to-buy-rival-randgold-in-all-share-deal

"Grand Inga Dam, DR Congo," *International Rivers.* www.internationalrivers.org/campaigns/grand-inga-dam-dr-congo

Hammond, M. "The Grand Ethiopian Renaissance Dam and the Blue Nile: Implications for transboundary water governance," *GWF Discussion Paper 1306,* Global Water Forum (2013). www.globalwaterforum.org/2013/01/25/the-grand-ethiopian-renaissance-dam-and-the-blue-nile-implications-for-transboundary-water-governance/

Negash, Fekahmed (Executive Director, Eastern Nile Technical Regional Office). Personal interview. 28 April 2018. Cambridge, MA.

Pearce, Fred. "On the River Nile, a Move to Avert a Conflict Over Water," *Yale Environment 360.* Yale School of Forestry & Environmental Studies. 12 Mar. 2015. e360.yale.edu/features/on_the_river_nile_a_move_to_avert_a_conflict_over_water

Stokstad, Erik. "Power play on the Nile," *Science* 351.6276 (2016): 904–907. doi.org/10.1126/science.351.6276.904

Taye, Meron Teferi, et al. "The Grand Ethiopian Renaissance Dam: Source of Cooperation or Contention?" *Journal of Water Resources Planning and Management* 142.11 (2016). doi.org/10.1061/(ASCE)WR.1943-5452.0000708

Three Inventions That Changed the World; Artificial Rivers; An Irishman Builds Los Angeles

Bagwell, Philip, and Peter Lyth. *Transport in Britain, 1750–2000: From Canal Lock to Gridlock* (London: Hambledon and London, 2002).

Davis, Mike. *City of Quartz: Excavating the Future in Los Angeles* (New York: Vintage Books, 2006).

Doyle, Martin. *The Source: How Rivers Made America and America Remade Its Rivers* (New York/London: W. W. Norton, 2018).

History of Canals in Britain—Routes of the Industrial Revolution. London Canal Museum. www.canalmuseum.org.uk/history/ukcanals.htm

Johnson, Ben. "The Bridgewater Canal," *Historic UK.* www.historic-uk.com/HistoryMagazine/DestinationsUK/The-Bridgewater-Canal/

Karas, Slawomir, and Maciej Roman Kowal. "The Mycenaean Bridges—Technical Evaluation Trial," *Roads and Bridges* 14.4 (2015): 285–302. doi.org/10.7409/rabdim.015.019

Redmount, Carol A. "The Wadi Tumilat and the 'Canal of the Pharaohs,'" *Journal of Near Eastern Studies* 54.2 (1995): 127–135. www.jstor.org/stable/545471

Wang, Serenitie, and Andrea Lo. "How the Nanjing Yangtze River Bridge Changed China Forever," *CNN Style: Architecture,* 2 Aug. 2017. www.cnn.com/style/article/nanjing-yangtze-river-bridge-revival

Wiseman, Ed. "Beipanjiang Bridge, the World's Highest, Opens to Traffic in Rural China," *The Telegraph,* 30 Dec. 2016. www.telegraph.co.uk/cars/news/beipanjiang-bridge-worlds-tallest-opens-traffic-rural-china/

Grand Diversions; Grand Bargains

Bagla, Pallava. "India plans the grandest of canal networks," *Science* 345.6193 (2014): 128. doi.org/10.1126/science.345.6193.128

Bhardwaj, Mayank. "Modi's $87 billion river-linking gamble set to take off as floods hit India," *Environment,* Reuters, 31 Aug. 2017. www.reuters.com/article/us-india-rivers/modis-87-billion-river-linking-gamble-set-to-take-off-as-floods-hit-india-idUSKCN1BC3HD

Bo, Xiang. "Water Diversion Project Success in First Year," *News,* English.news.cn. 12 Dec. 2015. www.xinhuanet.com/english/2015-12/12/c_134910168.htm

"From Congo Basin to Lake Chad: Transaqua, A Dream Is Becoming Reality," *Top News.* Sudanese Media Center. 29 Dec. 2016. smc.sd/en/from-congo-basin-to-lake-chad-transaqua-a-dream-is-becoming-reality/

Mekonnen, Mesfin M., and Arjen Y. Hoekstra. "Four billion people facing severe water scarcity," *Science Advances* 2.2 (2016). doi.org/10.1126/sciadv.1500323

Mengjie, ed. "South-to-north water diversion benefits 50 mln Chinese," *Xinhuanet,* 14 Sept. 2017. www.xinhuanet.com/english/2017-09/14/c_136609886.htm

"Metropolitan Board approves additional funding for full-scale, two-tunnel California WaterFix," The Metropolitan Water District of Southern California. 2 April 2018. www.mwdh2o.com/PDF_NewsRoom/WaterFix_April_board_decision.pdf

Mirza, Monirul M. Q., et al., eds. *Interlinking of Rivers in India: Issues and Concerns* (Leiden, Netherlands: CRC Press/Balkema, 2008).

Pateriya, Anupam. "Madhya Pradesh: Ken-Betwa river linking project runs into troubled waters," *Hindustan Times,* 8 July 2017. www.hindustantimes.com/india-news/madhya-pradesh-ken-betwa-river-linking-project-runs-into-troubled-waters/story-Sngb6U8mq2OeTMlB57KGsL.html

"Saving Lake Chad." *African Business Magazine,* 18 Apr. 2018. africanbusinessmagazine.com/sectors/development/saving-lake-chad

Whitehead, P. G., et al. "Dynamic modeling of the Ganga river system: impacts of future climate and socio-economic change on flows and nitrogen fluxes in India and Bangladesh," *Environmental Science: Processes & Impacts* 17 (2015): 1082–109. doi.org/10.1039/C4EM00616J

Zhao, Zhen-yu, et al. "Transformation of water resource management: a case study of the South-to-North Water Diversion project," *Journal of Cleaner Production* 163.1 (2017): 136–145. doi.org/10.1016/j.jclepro.2015.08.066

Chapter 6 — Pork Soup

A Super Fund

Beck, Eckardt C. "The Love Canal Tragedy," *EPA Journal*, Jan. 1979. archive.epa.gov/epa/aboutepa/love-canal-tragedy.html

Bedard, Paul. "Success: EPA set to reduce staff 50% in Trump's first term," *Washington Examiner*, 9 Jan. 2018. www.washingtonexaminer.com/success-epa-set-to-reduce-staff-50-in-trumps-first-term

Darland, Gary, et al. "A Thermophilic, Acidophilic Mycoplasma Isolated from a Coal Refuse Pile," *Science* 170.3965 (1970): 1416–1418. www.jstor.org/stable/1730880?seq=1/subjects

Davenport, Coral. "Scott Pruitt, Trump's Rule-Cutting E.P.A. Chief, Plots His Political Future," *New York Times*, 17 Mar. 2018. www.nytimes.com/2018/03/17/climate/scott-pruitt-political-ambitions.html

Deakin, Roger. *Waterlog: A Swimmer's Journey Through Britain* (New York: Vintage, 2000).

Guli, Mina. "The 6 River Run," www.minaguli.com/projectsoverview

———. "What I Learned from Running 40 Marathons in 40 Days for Water," *Huffpost*, 3 May 2017. www.huffpost.com/entry/what-i-learned-from-running-40-marathons-in-40-days_b_591acc92e4b03e1c81b008a1

Johnson, D. Barrie, and Kevin B. Hallberg. "Acid mine drainage remediation options: a review," *Science of the Total Environment* 338.1–2 (2005): 3–14. doi.org/10.1016/j.scitotenv.2004.09.002

Mallet, Victor. *River of Life, River of Death: The Ganges and India's Future* (New York: Oxford University Press, 2017).

Marsh, Rene. "Leaked memo: Pruitt taking control of Clean Water Act determinations," *CNN: Politics*, 4 April 2018. www.cnn.com/2018/04/04/politics/clean-water-act-epa-memo/index.html

O'Grady, John. "The $79 million plan to gut EPA staff," *The Hill*, 16 Feb. 2018. thehill.com/opinion/energy-environment/374167-the-79-million-plan-to-gut-epa-staff

Smith, L. C., and G. A. Olyphant. "Within-storm variations in runoff and sediment export from a rapidly eroding coal-refuse deposit," *Earth Surface Processes and Landforms* 19 (1994): 369–375. doi.org/10.1002/esp.3290190407

Swain, Christopher. "Swim with Swain." www.swimwithswain.org

China's River Chiefs; Streams of Malady

"China appoints 200,000 'river chiefs,'" *Xinhuanet*, 23 Aug. 2017. www.xinhuanet.com/english/2017-08/23/c_136549637.htm

Diaz, Robert J., and Rutger Rosenberg. "Spreading dead zones and consequences for marine ecosystems," *Science* 321.5891 (2008): 926–929. doi.org/10.1126/science.1156401

Eerkes-Medrano, Dafne, et al. "Microplastics in drinking water: A review and assessment," *Current Opinion in Environmental Science & Health* 7 (2019): 69–75. doi.org/10.1016/j.coesh.2018.12.001

Jensen-Cormier, Stephanie. "China Commits to Protecting the Yangtze River," *International Rivers,* 26 Feb. 2018. www.internationalrivers.org/blogs/435/china-commits-to-protecting-the-yangtze-river

Jones, Christopher S., et al. "Iowa stream nitrate and the Gulf of Mexico," *PLoS ONE* 13.4 (2018). doi.org/10.1371/journal.pone.0195930

Leung, Anna, et al. "Environmental contamination from electronic waste recycling at Guiyu, southeast China," *Journal of Material Cycles and Waste Management* 8.1 (2006): 21–33. doi.org/10.1007/s10163-005-0141-6

U.S. Fish and Wildlife Service, "Intersex fish: Endocrine disruption in smallmouth bass." www.fws.gov/chesapeakebay/pdf/endocrine.pdf

Williams, R. J., et al. "A national risk assessment for intersex in fish arising from steroid estrogens," *Environmental Toxicology and Chemistry* 28.1 (2009): 220–230. doi.org/10.1897/08-047.1

The Riviera of Greenland

Davenport, Coral, et al. "Greenland is melting away," *New York Times,* 27 Oct. 2015. www.nytimes.com/interactive/2015/10/27/world/greenland-is-melting-away.html

Fountain, H., and D. Watkins. "As Greenland Melts, Where Is the Water Going?" *New York Times,* 5 December 2017. www.nytimes.com/interactive/2017/12/05/climate/greenland-ice-melting.html

Kolbert, Elizabeth. "Greenland Is Melting," *New Yorker,* 24 October 2016. www.newyorker.com/magazine/2016/10/24/greenland-is-melting

Smith, L. C., et al. "Direct measurements of meltwater runoff on the Greenland Ice Sheet surface," *Proceedings of the National Academy of Sciences (PNAS)* 114.50 (2017): E10622-E10631. doi.org/10.1073/pnas.1707743114

Peak Water

Bolch, T., et al. "The State and Fate of Himalayan Glaciers," *Science* 336.6079 (2012): 310–314. science.sciencemag.org/content/336/6079/310

Dottori, Francesco, et al. "Increased human and economic losses from river flooding with anthropogenic warming," *Nature Climate Change* 8 (2018): 781–786. www.nature.com/articles/s41558-018-0257-z

Green, Fergus, and Richard Denniss. "Cutting with both arms of the scissors: the economic and political case for restrictive supply-side climate policies," *Climatic Change* 150.1–2 (2018): 73–87. doi.org/10.1007/s10584-018-2162-x

Hirabayashi, Yukiko, et al. "Global flood risk under climate change," *Nature Climate Change* 3 (2013): 816–821. www.nature.com/articles/nclimate1911

Huss, Matthias, and Regine Hock. "Global-scale hydrological response to future glacier mass loss," *Nature Climate Change* 8 (2018): 135–140. www.nature.com/articles/s41558-017-0049-x

Immerzeel, Walter W., et al. "Climate change will affect the Asian water towers," *Science* 328.5984 (2010): 1382–1385. science.sciencemag.org/content/328/5984/1382

Mallakpour, Iman, and Gabriele Villarini. "The changing nature of flooding across the central United States," *Nature Climate Change* 5 (2015): 250–254. www.nature.com/articles/nclimate2516

Milliman, J. D., et al. "Climatic and anthropogenic factors affecting river discharge to the global ocean, 1951–2000," *Global and Planetary Change* 62.3–4 (2008): 187–194. doi.org/10.1016/j.gloplacha.2008.03.001

Smith, Laurence C. "Trends in Russian Arctic river-ice formation and breakup: 1917 to 1994," *Physical Geography* 21.1 (2000): 46–56. doi.org/10.1080/02723646.2000.10642698

Udall, Bradley, and Jonathan Overpeck. "The twenty-first century Colorado River hot drought and implications for the future," *Water Resources Research* 53.3: 2404–2418. agupubs.onlinelibrary.wiley.com/doi/pdf/10.1002/2016WR019638

Woodhouse, Connie A., et al. "Increasing influence of air temperature on upper Colorado River streamflow," *Geophysical Research Letters* 43.5 (2016): 2174–2181. agupubs.onlinelibrary.wiley.com/doi/full/10.1002/2015GL067613

Xiao, Mu, et al. "On the causes of declining Colorado River streamflows," *Water Resources Research* 54.9 (2018): 6739–6756. agupubs.onlinelibrary.wiley.com/doi/abs/10.1029/2018WR023153

Chapter 7 — Going with the Flow

Dambusters Redux

American Rivers (2019): American River Dam Removal Database. Dataset www.americanrivers.org/2018/02/dam-removal-in-2017/

Foley, M. M., et al. "Dam removal: Listening in," *Water Resources Research* 53.7 (2017): 5229–5246. doi.org/10.1002/2017WR020457

Monterey Peninsula Water Management District, *San Clemente Dam Fish Counts.* www.mpwmd.net/environmental-stewardship/carmel-river-steelhead-resources/san-clemente-dam-fish-counts/

Schiermeier, Quirin. "Europe is demolishing its dams to restore ecosystems," *Nature* 557 (2018): 290–291. doi.org/10.1038/d41586-018-05182-1

Steinbeck, John. *Cannery Row* (New York: Penguin Group, Reprint Edition, 2002).

Williams, Thomas H., et al. "Removal of San Clemente Dam did more than restore fish passage," *The Osprey* 89 (2018): 1, 4–9. USGS Steelhead Committee Fly Fishers International. pubs.er.usgs.gov/publication/70195992

Starved for Sediment

Smith, Norman D., et al. "Anatomy of an avulsion," *Sedimentology* 36.1 (1989): 1–23. doi.org/10.1111/j.1365-3091.1989.tb00817.x

———. "Dam-induced and natural channel changes in the Saskatchewan River below the E. B. Campbell Dam, Canada," *Geomorphology* 260 (2016): 186–202. doi.org/10.1016/j.geomorph.2016.06.041

Smith, Norman D., and Marta Pérez-Arlucea. "Natural levee deposition during the 2005 flood of the Saskatchewan River," *Geomorphology* 101.4 (2008): 583–594. doi.org/10.1016/j.geomorph.2008.02.009

Smith, Norman D. et al. "Channel enlargement by avulsion-induced sediment starvation in the Saskatchewan River," *Geology* 42 (2014): 355–358. https://doi.org/10.1130/G35258.1.

Lessening the Damage

Chen, W., and J. D. Olden. "Designing flows to resolve human and environmental water needs in a dam-regulated river," *Nature Communications* 8.2158 (2017). doi.org/10.1038/s41467-017-02226-4

Holtgrieve, G. W., et al. "Response to Comments on "Designing river flows...," *Science* 13.6398 (2018). doi.org/10.1126/science.aat1477

Kondolf, G. Mathias, et al. "Dams on the Mekong: Cumulative sediment starvation," *Water Resource Research* 50.6 (2014): 5158–5169. doi.org/10.1002/2013WR014651

———. "Sustainable sediment management in reservoirs and regulated rivers: Experiences from five continents," *Earth's Future* 2 (2014): 256–280. doi.org/10.1002/2013EF000184

Sabo, J. L., et al. "Designing river flows to improve food security futures in the Lower Mekong Basin," *Science* 358.6368 (2017). doi.org/10.1126/science.aao1053

Schmitt, R. J. P., et al. "Improved trade-offs of hydropower and sand connectivity by strategic dam planning in the Mekong," *Nature Sustainability* 1 (2018): 96–104. www.nature.com/articles/s41893-018-0022-3

Zarfl, C., et al. "A global boom in hydropower dam construction," *Aquatic Sciences* 77.1 (2015): 161–170. doi.org/10.1007/s00027-014-0377-0

Wheels of the Future; Small Hydro in Big China

Harlan, Tyler. "Rural utility to low-carbon industry: Small hydropower and the industrialization of renewable energy in China," *Geoforum* 95 (2018): 59–69. doi.org/10.1016/j.geoforum.2018.06.025

Hennig, Thomas, and Tyler Harlan. "Shades of green energy: Geographies of small hydropower in Yunnan, China and the challenges of over-development," *Global Environmental Change* 49 (2018): 116–128. doi.org/10.1016/j.gloenvcha.2017.10.010

Khennas, Smail, and Andrew Barnett. *Best practices for sustainable development of micro hydro power in developing countries* (Department for International Development, UK: 2000). openknowledge.worldbank.org/handle/10986/20314

Low Impact Hydropower Institute, 2018. lowimpacthydro.org/wp-content/uploads/2018/05/2018LIHIFactSheet.pdf

Low Impact Hydropower Institute, "Pending Applications." lowimpacthydro.org/pending-applications-2/

Snakehead Stew; State-of-the-Art Salmon; Accidental Aquaculture

Brooks, A., et al. *A characterization of community fish refuge typologies in rice field fisheries ecosystems* (Penang, Malaysia: WorldFish, 2015).

Brown, David A. "Stop Asian Carp, Earn $1 Million," 2 Feb. 2017. www.outdoorlife.com/stop-asian-carp-earn-1-million/

Food and Agriculture Organization of the United Nations. *2018 The State of World Fisheries and Aquaculture* (2018). www.fao.org/state-of-fisheries-aquaculture/en/

Ge, Celine. "China's Craving for Crayfish Creates US$2 Billion Business," *South China Morning Post,* 26 June 2017. www.scmp.com/business/companies/article/2100001/chinas-craving-crayfish-creates-us2-billion-business

Love, Joseph W., and Joshua J. Newhard. "Expansion of Northern Snakehead in the Chesapeake Bay Watershed," *Transactions of the American Fisheries Society* 147.2 (2018): 342–349. doi.org/10.1002/tafs.10033

Lu, Lula, and John Crilley (founders and co-owners, FIn Gourmet Foods). Personal interview. Teleconference, 29 Aug. 2018. Paducah, KY.

Penn, Ivan. "The $3 Billion Plan to Turn Hoover Dam into a Giant Battery," *New York Times,* 24 July 2018. www.nytimes.com/interactive/2018/07/24/business/energy-environment/hoover-dam-renewable-energy.html

Rehman, Shafiqur, et al. "Pumped hydro energy storage system: A technological review," *Renewable and Sustainable Energy Reviews* 44 (2015): 586–598. doi.org/10.1016/j.rser.2014.12.040

Souty-Grosset, Catherine, et al. "The red swamp crayfish *Procambarus clarkii* in Europe: Impacts on aquatic ecosystems and human well-being," *Limnologica* 58 (2016): 78–93. doi.org/10.1016/j.limno.2016.03.003

New Roles for Old Rivers; The Three-Billion-Dollar Battery; Empty Your Bowl; A Dark Desert Highway

Calico, Rachel (USACE), Caitlin Campbell (USACE), Angela DeSoto (Jefferson Parish), and Larry Nulty (Pump Station Superintendent). Personal Interviews. 14 Dec. 2017. SELA "Pump to the River" pumping station at 1088 Dickory, Jefferson, LA.

Maloney, Peter. "Los Angeles Considers $3B Pumped Storage Project at Hoover Dam," *Utility Dive,* 26 July 2018. www.utilitydive.com/news/los-angeles-considers-3b-pumped-storage-project-at-hoover-dam/528699/

Metropolitan Water District of Southern California. "Metropolitan study demonstrates feasibility of large-scale regional recycling water program," 9 January 2017. www.mwdh2o.com/PDF_NewsRoom/RRWP_FeasibilityStudyRelease.pdf

Kightlinger, Jeffrey. Personal Interview, 16 Sept. 2017, at the Metropolitan Headquarters in Los Angeles, California. Other Kightlinger heard at the Groundbreaking Ceremony for the Regional Recycled Water Advanced Purification Center, 18 Sept. 2017, at the Joint Water Pollution Control Plant (JWPCP), Carson, California.

"Southeast Louisiana Urban Flood Control Project—SELA," U.S. Army Corps of

Engineers, Feb. 2018. www.mvn.usace.army.mil/Portals/56/docs/SELA/SELA Fact Sheet Feb 2018.pdf. rsc.usace.army.mil/sites/default/files/MR&T_17 Jun15_Final.pdf

Wehner, Michael. Personal Interview, 20 Sept. 2019, at the Orange County Water District GWRS (Groundwater Replenishment System), Fountain Valley, California.

Chapter 8 — A Thirst for Data

The River's Purpose

Blom, A., et al. "The graded alluvial river: Profile concavity and downstream fining," *Geophysical Research Letters* 43.12 (2016): 6285–6293. doi.org/10.1002/2016GL068898

Cassis, N. "Alberto Behar (1967–2015)," *Eos* 96 (2015). doi.org/10.1029/2015EO032047

Mackin, J. H. "Concept of the Graded River," *Bulletin of the Geological Society of America* 59 (1948): 463–512. doi.org/10.1177/030913330002400405

Tireless Toil vs. Fire and Ice

Magilligan, F. J., et al. "Geomorphic effectiveness, sandur development and the pattern of landscape response during jökulhlaups: Skeiðarársandur, southeastern Iceland," *Geomorphology* 44.1–2 (2002): 95–113. doi.org/10.1016/S0169-555X(01)00147-7

Smith, L. C., et al. "Estimation of erosion, deposition, and net volumetric change caused by the 1996 Skeiðarársandur jökulhlaup, Iceland, from Synthetic Aperture Radar Interferometry," *Water Resources Research* 36.6 (2000): 1583–1594. doi.org/10.1029/1999WR900335

———. "Geomorphic impact and rapid subsequent recovery from the 1996 Skeiðarársandur jökulhlaup, Iceland, assessed with multi-year airborne lidar," *Geomorphology* 75 (2006): 65–75. doi.org/10.1016/j.geomorph.2004.01.012

Documentarians of Earth

Allen, G. H., et al. "Similarity of stream width distributions across headwater systems," *Nature Communications* 9.610 (2018). doi.org/10.1038/s41467-018-02991-w

Allen, G. H., and T. M. Pavelsky. "Global extent of rivers and streams," *Science* 361.6402 (2018): 585–588. doi.org/10.1126/science.aat0636

Cooley, S. W., et al. "Tracking Dynamic Northern Surface Water Changes with High-Frequency Planet CubeSat Imagery," *Remote Sensing* 9.12 (2017): 1306. doi.org/10.3390/rs9121306

Fishman, Charles. *One Giant Leap* (New York: Simon & Schuster, 2019).

Gleason, Colin J., and Laurence C. Smith. "Toward global mapping of river discharge using satellite images and at-many-stations hydraulic geometry,"

Proceedings of the National Academy of Sciences (PNAS) 111.13 (2014): 4788–4791. doi.org/10.1073/pnas.1317606111

Gleason, C. J., et al. "Retrieval of river discharge solely from satellite imagery and at-many-stations hydraulic geometry: Sensitivity to river form and optimization parameters," *Water Resources Research* 50 (2014): 9604–9619. doi.org/10.1002/2014WR016109

Pekel, J.-F., et al. "High-resolution mapping of global surface water and its long-term changes," *Nature* 540 (2016): 418–422. doi.org/10.1038/nature20584

Smith, L. C., and T. M. Pavelsky. "Estimation of river discharge, propagation speed, and hydraulic geometry from space: Lena River, Siberia," *Water Resources Research* 44.3 (2008): W03427. doi.org/10.1029/2007WR006133

Put on Your 3-D Glasses; Big Data Meets Global Water; The Power of Models

Alsdorf, D. E., et al. "Amazon water level changes measured with interferometric SIR-C radar," *IEEE Transactions on Geoscience and Remote Sensing* 39.2 (2001): 423–431. doi.org/10.1109/36.905250

———. "Interferometric radar measurements of water level change on the Amazon flood plain," *Nature* 404 (2000): 174–177. doi.org/10.1038/35004560

———. "Spatial and temporal complexity of the Amazon flood measured from space," *Geophysical Research Letters* 34 (2007): L080402. doi.org/10.1029/2007GL029447

Altenau, Elizabeth H., et al. "AirSWOT measurements of river water surface elevation and slope: Tanana River, AK," *Geophysical Research Letters* 44 (2017): 181–189. doi.org/10.1002/2016GL071577

Biancamaria, S., et al. "The SWOT Mission and Its Capabilities for Land Hydrology," *Surveys in Geophysics* 37.2 (2016): 307–337. doi.org/10.1007/s10712-015-9346-y

Deming, D. "Pierre Perrault, the Hydrologic Cycle and the Scientific Revolution," *Groundwater* 152.1 (2014): 156–162. doi.org/10.1111/gwat.12138

Pavelsky, Tamlin M., et al. "Assessing the potential global extent of SWOT river discharge observations," *Journal of Hydrology* 519, Part B (2014): 1516–1525. doi.org/10.1016/j.jhydrol.2014.08.044

Pitcher, Lincoln H., et al. "AirSWOT InSAR Mapping of Surface Water Elevations and Hydraulic Gradients Across the Yukon Flats Basin, Alaska," *Water Resources Research* 55.2 (2019): 937–953. doi.org/10.1029/2018WR023274

Rodríguez, Ernesto, et al. "A Global Assessment of the SRTM Performance," *Photogrammetric Engineering & Remote Sensing* 3 (2006): 249–260. doi.org/10.14358/PERS.72.3.249

Shiklomanov, I. A. "World Fresh Water Resources," in P. H. Gleick, ed., *Water in Crisis* (New York: Oxford University Press, 1993), 13–24.

Chapter 9 — Rivers Rediscovered

An Unnatural Separation

Kesebir, S., and Pelin Kesebir. "A Growing Disconnection from Nature Is Evident in Cultural Products," *Perspectives on Psychological Science* 12.2 (2017): 258–269. doi.org/10.1177/1745691616662473

New York State Department of Environmental Conservation, Fish and Wildlife, 625 Broadway, Albany, NY 12233-4754.

Pergams, Oliver R. W., and Patricia Zaradic. "Evidence for a fundamental and pervasive shift away from nature-based recreation," *Proceedings of the National Academy of Sciences (PNAS)* 105.7 (2008): 2295–2300. doi.org/10.1073/pnas.0709893105

Prévot-Julliard, A.-C., et al. "Historical evidence for nature disconnection in a 70-year time series of Disney animated films," *Public Understanding of Science* 24.6 (2015): 672–680. doi.org/10.1177/0963662513519042

Price Tack, Jennifer L., et al. "Managing the vanishing North American hunter: a novel framework to address declines in hunters and hunter-generated conservation funds," *Human Dimensions of Wildlife* 23.6 (2018): 515–532. doi.org/10.1080/10871209.2018.1499155

U.S. Fish and Wildlife Service, "Historical License Data." wsfrprograms.fws.gov/Subpages/LicenseInfo/LicenseIndex.htm

Zaradic, Patricia, and Oliver R. W. Pergams, "Trends in Nature Recreation: Causes and Consequences," *Encyclopedia of Biodiversity* 7 (2013): 241–257. doi.org/10.1016/B978-0-12-384719-5.00321-X

Nature and the Brain

Barton, Jo, and Jules Pretty. "What is the best dose of nature and green exercise for improving mental health? A multi-study analysis," *Environmental Science and Technology* 44.10 (2010): 3947–3955. doi.org/10.1021/es903183r

Barton, Jo, Murray Griffin, and Jules Pretty. "Exercise-, nature- and socially interactive-based initiatives improve mood and self-esteem in the clinical population," *Perspectives in Public Health* 132.2 (2012): 89–96. doi.org/10.1177/1757913910393862

Berman, M. G., et al. "The Cognitive Benefits of Interacting with Nature," *Psychological Science* 19.12 (2008): 1207–1212. doi.org/10.1111/j.1467-9280.2008.02225.x

Brown, Adam, Natalie Djohari, and Paul Stolk. *Fishing for Answers: The Final Report of the Social and Community Benefits of Angling Project* (Manchester, UK: Substance, 2012). resources.anglingresearch.org.uk/project_reports/final_report_2012

Freeman, Claire, and Yolanda Van Heezik. *Children, Nature and Cities* (London: Routledge, 2018).

Kuo, M. "How might contact with nature promote human health? Promising mechanisms and a possible central pathway," *Frontiers in Psychology* 25 (2015): 1–8. doi.org/10.3389/fpsyg.2015.01093

Louv, Richard. *Last Child in the Woods* (Chapel Hill, NC: Algonquin Books of Chapel Hill, 2008).

———. *The Nature Principle: Reconnecting with Life in a Virtual Age* (Chapel Hill, NC: Algonquin Books of Chapel Hill, Reprint Edition, 2012).

Nichols, Wallace J. *Blue Mind* (New York: Little, Brown and Company, 2015).

Three Moments in Manhattan

Beauregard, Natalie. "High Line Architects Turn Historic Brooklyn Sugar Factory into Sweet Riverside Park," *AFAR*, 6 June 2018. www.afar.com/magazine/a-riverfront-park-grows-in-brooklyn

"History of the Yard." *A Place to Build Your History*, Brooklyn Navy Yard. brooklynnavyyard.org/about/history

Kimball, A. H., and D. Romano. "Reinventing the Brooklyn Navy Yard: a national model for sustainable urban industrial job creation," *WIT Transaction on the Built Environment* 123 (2012): 199–206. doi.org/10.2495/DSHF120161

Pulver, Liz (Landscape Architect PLLC, Liz Pulver Design). Personal Interview. 25 March 2018. Brooklyn, NY.

Tamir, Karen (James Corner Field Operations), and Jovana Rizzo (Berlin Rosen). Personal Interviews. 11 May 2018. Greenpoint Landing, Brooklyn, NY.

Global Urban Renewals; The Urban Majority

Barragan, Bianca. "Mapped: 21 Projects Rising along the LA River," *Curbed Los Angeles*, 3 May 2018. la.curbed.com/maps/los-angeles-river-development-map-sixth-street-bridge

"Battersea Power Station." Battersea Power Station Iconic Living. batterseapowerstation.co.uk

"Brooklyn Bridge Plaza." Brooklyn Bridge Park. www.brooklynbridgepark.org/pages/futurepark

Chiland, Elijah. "New Plans Could Reshape 19 Miles of the LA River, from Vernon to Long Beach," *Curbed Los Angeles*, 14 Dec. 2017. la.curbed.com/2017/12/14/16776934/la-river-plans-revitalization-vernon-long-beach

Cusack, Brennan. "Egypt Is Building Africa's Tallest Building," *Forbes*, 28 Aug. 2018. www.forbes.com/sites/brennancusack/2018/08/28/egypt-is-building-africas-tallest-building/#3ead57512912

da Fonseca-Wollheim, Corinna, "Finally, a Debut for the Elbphilharmonie Hall in Hamburg," *New York Times*, 10 Jan. 2017. www.nytimes.com/2017/01/10/arts/music/elbphilharmonie-an-architectural-gift-to-gritty-hamburg-germany.html

Garcetti, Eric. "Mayor Garcetti Celebrates Final Acquisition of Land Considered 'Crown Jewel' in Vision to Revitalize L.A. River," Mayor Eric Garcetti, City of Los Angeles, 3 Mar. 2017. www.lamayor.org/mayor-garcetti-celebrates-final-acquisition-land-considered-%e2%80%98crown-jewel%e2%80%99-vision-revitalize-la-river

Garfield, Leanna. "6 Billion-Dollar Projects That Will Transform London by 2025," *Business Insider,* 22 Aug. 2017. www.businessinsider.com/london-mega projects-that-will-transform-the-city-2017-8

"Mott Haven–Port Morris Waterfront Plan." South Bronx Unite. southbronx unite.org/a-waterfront-re-envisioned

"New York City Comprehensive Waterfront Plan," *Vision 2020,* NYC Department of City Planning. www1.nyc.gov/site/planning/plans/vision-2020-cwp/vision -2020-cwp.page

United Nations Department of Economic and Social Affairs. "2018 Revision of World Urbanization Prospects," Publications, www.un.org/development/desa /publications/2018-revision-of-world-urbanization-prospects.html

White, Anna, "Exclusive First Look: London's 4.2bn Pound Thames Tideway Super Sewer Is an Unprecedented Planning Victory to Build into the River," *Homes & Property,* 4 Sept. 2018. www.homesandproperty.co.uk/property -news/buying/new-homes/londons-new-super-sewer-to-open-up-the-thames -with-acres-of-public-space-for-watersports-arts-and-a123641.html

INDEX

ABOUT THE AUTHOR

Laurence C. Smith is the John Atwater and Diana Nelson University Professor of Environmental Studies and professor of Earth, Environmental and Planetary Sciences at Brown University. Previously, he was professor and chair of Geography at the University of California, Los Angeles. A fellow of the American Geophysical Union and of the John S. Guggenheim Foundation, his first book, *The World in 2050,* was winner of the Walter P. Kistler Book Award and a *Nature* Editor's Pick of 2012. He is a frequent public speaker, including invited keynotes at the World Economic Forum in Davos. His scientific research has been featured in the *New York Times,* the *Wall Street Journal, The Economist,* the *Los Angeles Times,* the *Washington Post, The Globe and Mail,* the *Financial Times,* and *Discover Magazine* and on NPR, CBC Radio, and the BBC, among others.